AF709293

BIOTECHNOLOGY IN AGRICULTURE, INDUSTRY AND MEDICINE

CLOSTRIDIA

BIOTECHNOLOGY, MEDICINAL APPLICATIONS AND IMPLICATIONS

BIOTECHNOLOGY IN AGRICULTURE, INDUSTRY AND MEDICINE

Additional books in this series can be found on Nova's website under the Series tab.

Additional E-books in this series can be found on Nova's website under the E-books tab.

BACTERIOLOGY RESEARCH DEVELOPMENTS

Additional books in this series can be found on Nova's website under the Series tab.

Additional E-books in this series can be found on Nova's website under the E-books tab.

BIOTECHNOLOGY IN AGRICULTURE, INDUSTRY AND MEDICINE

CLOSTRIDIA

BIOTECHNOLOGY, MEDICINAL APPLICATIONS AND IMPLICATIONS

STEVEN V. CLINTON
AND
MICHAEL F. KELLY
EDITORS

Nova Science Publishers, Inc.
New York

Copyright © 2012 by Nova Science Publishers, Inc.

All rights reserved. No part of this book may be reproduced, stored in a retrieval system or transmitted in any form or by any means: electronic, electrostatic, magnetic, tape, mechanical photocopying, recording or otherwise without the written permission of the Publisher.

For permission to use material from this book please contact us:
Telephone 631-231-7269; Fax 631-231-8175
Web Site: http://www.novapublishers.com

NOTICE TO THE READER

The Publisher has taken reasonable care in the preparation of this book, but makes no expressed or implied warranty of any kind and assumes no responsibility for any errors or omissions. No liability is assumed for incidental or consequential damages in connection with or arising out of information contained in this book. The Publisher shall not be liable for any special, consequential, or exemplary damages resulting, in whole or in part, from the readers' use of, or reliance upon, this material. Any parts of this book based on government reports are so indicated and copyright is claimed for those parts to the extent applicable to compilations of such works.

Independent verification should be sought for any data, advice or recommendations contained in this book. In addition, no responsibility is assumed by the publisher for any injury and/or damage to persons or property arising from any methods, products, instructions, ideas or otherwise contained in this publication.

This publication is designed to provide accurate and authoritative information with regard to the subject matter covered herein. It is sold with the clear understanding that the Publisher is not engaged in rendering legal or any other professional services. If legal or any other expert assistance is required, the services of a competent person should be sought. FROM A DECLARATION OF PARTICIPANTS JOINTLY ADOPTED BY A COMMITTEE OF THE AMERICAN BAR ASSOCIATION AND A COMMITTEE OF PUBLISHERS.

Additional color graphics may be available in the e-book version of this book.

Library of Congress Cataloging-in-Publication Data

Clostridia : biotechnology, medicinal applications, and implications / editors, Steven V. Clinton and Michael F. Kelly.
p. ; cm.
Includes bibliographical references and index.
ISBN 978-1-62100-761-6 (hardcover)
I. Clinton, Steven V. II. Kelly, Michael F. (Michael Frances), 1972-
[DNLM: 1. Clostridium. 2. Biomedical Technology--methods. 3. Biotechnology--methods. QW 127.5.C5]

614.5'12--dc23

2011038323

Published by Nova Science Publishers, Inc. † New York

CONTENTS

PREFACE

The genus Clostridium consists of over 295 species of gram-positive, rod-shaped bacteria. These are mostly anaerobes and present ubiquitously in the environment due to their ability to form heat, radiation and chemical resistant spores. In this book, the authors present current research in the study of the biotechnology, medicinal applications and implications of clostridia. Topics discussed include the biodegradation of organic pollutants, reduction of metals and energy generation involving Clostridium sp.; the utilization of clostridium in industrial, biotechnological and medical endeavors; the clinical aspects of clostridium difficile associated with diarrhea and the potential use of megasphaera elsdenii as a hydrogen producer.

Chapter 1- The role of the genus *Clostridium* is of upmost importance in environmental biotechnology; the great metabolic diversity that the members of this genus exhibit allows them to participate in biodegradation of organic pollutants, biotransformation of metals and generation of energy.

In this review, the following topics will be presented: biodegradation of some relevant organic pollutants such as Azo-dyes, explosives, chlorinated compounds, biotransformation of metals such as Chromium, Vanadium, Iron, Uranium and Plutonium, and the generation of energy in Microbial Fuel Cells (MFCs) including the main metabolic features of some members of the *Clostridium* genus involved in these processes. Therefore, some key points in this review include:

Chapter 2- The genus *Clostridium* consists of over 295species of gram-positive, rod shaped bacteria. These are mostly strict anaerobes and present ubiquitously in the environment due to their ability to form heat, radiation and chemical resistant spores. A number of clostridial species also inhabit the gastrointestinal tracts of animals and humans. Most of clostridial species are harmless microbes; some even are well-known for being employed to benefit humans,but only a few cause diseases that gave the genus a bed name.

Chapter 3- *Clostridium difficile* is an important healthcare associated infection, in terms of incidence, associated morbidity and mortality, and economic impact. The recent emergence of a more virulent strain, ribotype 027, has resulted in an increased incidence of *Clostridium difficile*-associated diarrhoea (CDAD) across the world, with an attendant rise in the severity of cases. This article will review the diagnosis, treatment, and control of the disease, and discuss new and emerging management strategies.

Chapter 4- *Clostridium difficile* (*C.difficile*) is a gram-positive, spore-forming organism responsible for a spectrum of infections. These infections vary from asymptomatic colonization to severe antibiotic-related diarrhea and life threatening pseudomembranous colitis *C.difficile* forms part of the gastrointestinal tracts flora of most neonates and was believed to be a commensal organism until it was later discovered that *Clostridium difficile* is the primary cause of pseudomembranous colitis and the frequent use of broad-spectrum antibiotics lead to the rapid incidence of *C.difficile* diarrhea.

Chapter 5- Over the years, global energy consumption, particularly of fossil fuels, continues to increase. The planet's accessible oil reservoirs have been predicted to be almost depleted within this century. Hydrogen is a promising alternative fuel for sustainable use since the development of alternative energy sources is necessary to establish the long-term sustainability of modern societies. Microbial hydrogen production may reduce the dependence on fossil fuels, reduce carbon dioxide emissions, and unlock unused reserves of bioenergy. Hydrogen fermentation by anaerobic bacteria shows the greatest capacity for microbial hydrogen production. Within the field of hydrogen fermentation research, innovations from a microbiological perspective have been developing continually. In most hydrogen fermentation studies, the genus *Clostridium* has been used as the main hydrogen producer because it appears to produce abundant hydrogen. However, to date, very few industrial systems have progressed beyond the pilot study stage. This commentary will describe some of the problems associated with the development of industrial-level hydrogen production and will introduce *Megasphaera elsdenii* as one of the microbiological tools currently in development to solve these problems. *M. elsdenii* has some unique characteristics, including its high hydrogen productivity as hydrogen-producing bacteria (HPB) and lactate consumption as lactate-utilizing bacteria (LUB), which will be useful in the design of a practical hydrogen fermentation system.

In: Clostridia ISBN: 978-1-62100-761-6
Editors: S. V. Clinton and M. F. Kelly © 2012 Nova Science Publishers, Inc.

Chapter 1

BIODEGRADATION OF ORGANIC POLLUTANTS, REDUCTION OF METALS AND ENERGY GENERATION INVOLVING *CLOSTRIDIUM SP.*

Claudia Guerrero-Barajas
Departamento de Bioprocesos, Unidad Profesional Interdisciplinaria de Biotecnología, Instituto Politécnico Nacional, Mexico City, Mexico

ABSTRACT

The role of the genus *Clostridium* is of upmost importance in environmental biotechnology; the great metabolic diversity that the members of this genus exhibit allows them to participate in biodegradation of organic pollutants, biotransformation of metals and generation of energy.

In this review, the following topics will be presented: biodegradation of some relevant organic pollutants such as Azo-dyes, explosives, chlorinated compounds, biotransformation of metals such as Chromium, Vanadium, Iron, Uranium and Plutonium, and the generation of energy in Microbial Fuel Cells (MFCs) including the main metabolic features of some members of the *Clostridium* genus involved in these processes. Therefore, some key points in this review include:

The Azo-dyes (synthetic dyes) compounds, which are widely used in industry and can be biodegraded under anaerobic conditions by azoreductases, enzymes expressed by some *Clostridium* sp. A singular dehydrogenase identified in the case of *Clostridium perfringens* allows to this microorganism the reduction of nitro aromatic compounds as well.

The anaerobic biodegradation of nitrogenated compounds including the explosives 2,4,6- trinitrotoluene (TNT), hexahydro- 1,3,5, trinitro-1,3,5 triazine (RDX) and octahydro-1,3,5,7- tetranitro-1,3,5,7-tetrazocine (HMX) are linked to *Clostridium nitrophenolicum*, which is able to survive in mixtures of the mentioned explosives.

Among the several recalcitrant organic pollutants that can be biodegraded by microorganisms belonging to the genus *Clostridium* are also the chlorinated aliphatic and aromatic hydrocarbons, in which case *Clostridium bifermentans* DPH-1 is known to be capable of dehalorespiration. Successful results on polychlorinated biphenyls (PCBs) dechlorination involve *Clostridium pascui*.

Reduction of metals and radionuclids is another capability of Clostridia. The mechanisms for these reductions are not completely known in some cases, and both fermentation and respiration have been pointed out for the reductive process. Recently, it has been proposed that *Clostridium acetobutylicum* may reduce U(VI) enzymatically. Some of the most cited metals and radionuclids that several *Clostridium* sp. can reduce are Fe (III), by *Clostridium celerecrescens*, V(V), by *Clostridium pasteurianum* and U(VI) by *Clostridium acetobutylicum*.

A novel strategy for energy generation taking advantage of microorganisms is the microbial fuel cell (MFC). Some of the most studied microorganisms used in microbial fuel cells (MFCs) include *Clostridium butyricum* and *Clostridium acetobutylicum*. Most recently, an interesting application of MFCs is to couple the energy generation to anaerobic biodegradation of hydrocarbons, a process that also involves several microorganisms belonging to the genus *Clostridium*.

Certainly, the genus *Clostridium* offers a great source of bacteria that ensures a physiology complex enough to be taken advantage of in environmental biotechnology; however, there are still places to look at in which *Clostridium* sp. can be found and therefore, capabilities to be discovered. This may be true when exploring microbial niches in extreme environments such as the Antarctica, where a microorganism closely related to *Clostridium subterminale* exhibited the capability of producing an extracellular protease, active in a wide range of pH.

On the other hand, there is relatively little information of *Clostridium* sp. in marine environments around the world, and this is particularly true in thermophilic marine environments such as hydrothermal vents, for which only two *Clostridium* species have been reported: *Clostridium caminithermale and Clostridium tepidiprofundi*. Along these lines, a microorganism closely related to *Clostridium ganghwense* has been related to biodegradation of trichloroethylene (TCE) when enrichment cultures of hydrothermal vents sediments were cultivated under sulfate-reducing conditions.

Finally, in sight of the valuable biological tool that this genus is, emerging new contaminants such as "green solvents" have also been evaluated on the toxicity that may exert on *Clostridium* sp.; this is recent, and may become a great concern in a near future, since for the scientific community dedicated to biodegradation and bioremediation, the fact that some members of *Clostridium* are ubiquitous is definitely useful.

INTRODUCTION

The presence of organic pollutants and metals in soils and water streams is a problem that affects, to a different degree, a great extension of the inhabited planet. Nowadays, microbial processes that help to remove the pollution originated by this anthropogenic contamination are still studied to clarify the mechanisms that drive the biodegradation or biotransformation of the toxic compounds. Also, these microbial processes are often applied successfully to remediate contaminated sites.

Among the several groups of organic pollutants that have been biodegraded or biotransformed by microorganisms are, for example, the halogenated compounds, for which the chlorinated compounds have received great attention due to the importance that they pose to the human activities.

Explosives and their intermediates cannot be excluded from this brief list because they are still of great concern for many countries, either as a result of the manufacturing process or the final use that they are given. Another important group of organic nitrogenated pollutants is constituted by the Azo-dyes compounds, for which also a vast research has been conducted in order to remove their presence from soils and water streams.

In many cases, pollution by organic compounds is also accompanied by certain metals, in which case the microbial processes may become more complicated and, therefore, is more difficult to describe the steps that lead to decontamination of the site. Fortunately, in either case, microorganisms almost always present a strategy to survive, and biodegradation or biotransformation do occur along with the still difficult task of investigating what actually happened during the process. Enzymes involved, electrons flow, and the role of each contaminant are some of the key points to know in every biodegradation that occurs. Specific data such as kinetics parameters and composition of the microbial communities are currently available for biodegradation or biotransformation of several compounds belonging to the mentioned groups of pollutants and are always desirable in order to use the microbial process in engineered remediation.

Among the ubiquitous microorganisms that have demonstrated capability of biodegrading a wide variety of compounds, the genus *Clostridium* is one of the most cited.

The genus *Clostridium* is one of the most studied among the genera of prokaryotes that include anaerobic, gram-positive endospore-forming bacteria (or spore-forming bacteria). *Clostridium* sp. are bacteria largely known in environmental biotechnology due to their active and useful enzymes (i.e., nitroreductases), which have been studied in biodegradation and bioremediation of several highly toxic organic pollutants.

Besides the role that these bacteria may play during the biotransformation of certain pollutants, they have also been studied in systems designed to produce energy such as the relatively new Microbial Fuel Cells (MFCs).

Not only are they investigated in biodegradation and generation of energy, but also some strains identified in remote and relatively unexplored places may offer a source of biological compounds that may be of biotechnological interest; therefore, there are ongoing attempts to isolate and identify "new" *Clostridium* sp. around different places in the world.

The main objective of this chapter is to review the literature on the most recent information about the involvement of *Clostridium* sp. in biodegradation and biotransformation of pollutants and the role they play in MFCs. The groups of contaminants presented here were chosen based on the reports on *Clostridium* sp. involved in their biodegradation rather than the any other possible reason.

The chapter is organized as follows:

1) Biodegradation of organic pollutants and *Clostridium* sp. This section will cover briefly the source of each pollutant, the conditions reported for biodegradation or biotransformation to occur and the strain of *Clostridium* involved in the process.
2) Biotransformation of metals involving *Clostridium* sp. This section will cover the source of metals and radionuclides for which biological transformation has been linked to *Clostridium* sp. and the conditions reported for the biotransformation to occur as well as the strain of *Clostridium* involved. The mechanisms involved in the microbial processes covered in sections **1)** and **2)** will be described as detailed as possible, if there is enough information involving *Clostridium*.

3) Application of *Clostridium* sp. in MFCs. This section will describe briefly the components of an MFC and will mention some of the most cited cases in which *Clostridium* has been used in this type of system.

Finally, the last section of the chapter will present an outline of the potentially novel applications of certain *Clostridium* found in extreme or relatively unexplored environments or within specific biodegradation processes.

Should the reader need more specific details on the mechanisms involved in the microbial processes that will be described here, he may want to refer to the original reference since often, the description of the proteins and the electron chain transport will not be entirely shown.

BIODEGRADATION OF ORGANIC POLLUTANTS BY *CLOSTRIDIUM* SP.

Clostridium Sp. and Biodegradation of Azo-Dyes

One of the most important groups of organic pollutants is constituted by the Azo dyes compounds. Due to the great environmental hazard that these compounds are, there is information available on physicochemical and biological removal processes updated to the current year. There are some extensive and recent reviews on bacterial decolourization and degradation of azo dyes to which the reader may want to refer for details on the various mechanisms of degradation of these compounds. Two of the most recent review documents on the topic include one written by Pandey et al. (2007) and, to the best of our knowledge, the most recent published is the written by Saratale et al. (2011).

Azo dyes (-N=N-) are utilized in several industries such as textile, pharmaceutical, food, cosmetic and paper printing (Forgacs et al., 2004; Pandey et al., 2007). According to the existing reports on dyes research, this group of synthetic compounds is widely utilized in the textile industry, for which two-thirds of the total dye market is composed (Forgacs et al., 2004; Elisangela et al., 2009). One of the most recent documents providing the amount of azo dyes released to the environment is the one issued by Saratale et al. (2011), stating that approximately 280,000 t of textile dyes are discharged through industrial waters worldwide every year (Jin et al., 2007; Saratale et al., 2011).

The presence of members of this group of compounds is undesirable in water and wastewater streams where they usually find their final fate due to the large damage that they may cause to aquatic life. It is known that they and some of the intermediates resulting from their biological transformation can be classified as carcinogenic or mutagenic (Forgacs et al., 2004; Pandey et al., 2007; Saratale et al., 2011). In addition, they exert an adverse impact to total organic carbon (TOC), biological oxygen demand (BOD) and chemical oxygen demand (COD) (Saratale et al., 2011).

Azo dyes can be degraded under aerobic and anaerobic conditions, depending on the number of (-N=N-) substituents in the molecule. Figure 1 shows the structure of some Azo-dyes widely used in industry. Given the fact that this review covers aspects on *Clostridium*, only anaerobic conditions for Azo-dyes biodegradation will be mentioned.

Nowadays, it is well known that biological degradation of these compounds can be carried out by fungi, yeast, algae, plants and bacteria (Saratale et al., 2011). Details on the many factors such as pH, temperature and concentration of the dye that exert an effect on bacterial degradation of these compounds are described by Saratale et al. (2011).

It is known that anaerobic, bacterial biodegradation of these compounds can be (Pandey et al., 2007):

a) Enzymatic, in which case the specific enzymatic first step in bacterial degradation is the reduction of the azo bond (-N=N-)
b) Non-specific enzymatic, even though non-specific biodegradation of these compounds can occur under methanogenic conditions, in order to highlight the role of *Clostridium* sp. in the microbiological processes, only some aspects on the anaerobic processes will be mentioned in which *Clostridium* sp. have been demonstrated to play an important role.
c) Redox mediated Azo-dye reduction, in which another compound serves as electron shuttle for the reduction of the pollutant. Particularly in the case of Azo-dyes, this mechanism has been proved with anaerobic sludge and other consortia different from *Clostridium* (Van der Zee et al., 2001; Field and Brady, 2003).

A scheme of the above-mentioned mechanisms is briefly described in Figure 2.

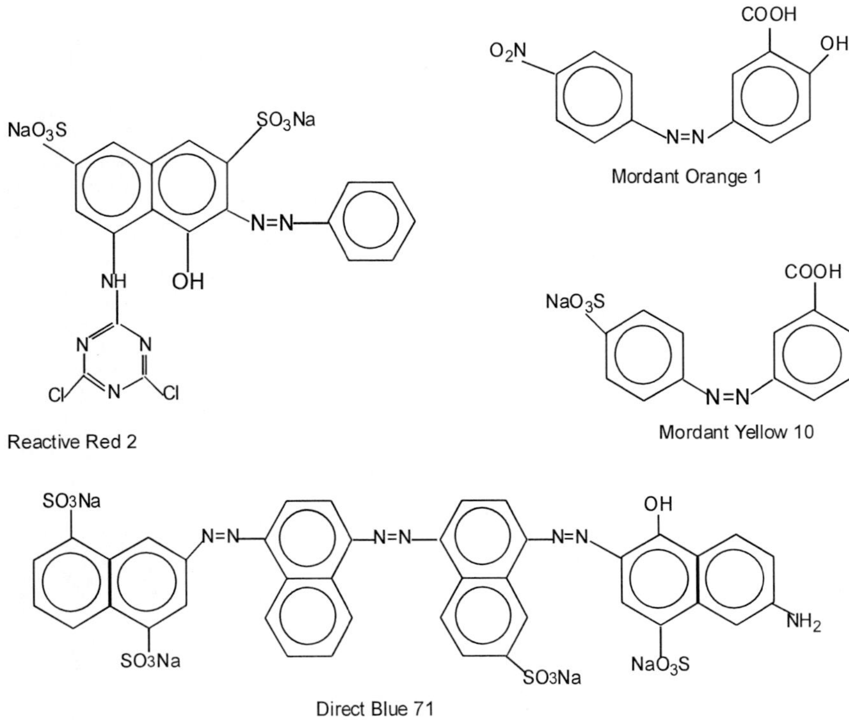

Figure 1. Chemical structures of some representative azo-dyes.

Under anaerobic conditions, direct, enzymatic degradation of azo dyes can occur in the presence and release of the oxygen sensitive azoreductases, which has been confirmed in the genera *Clostridium* and *Eubacterium*. (Pandey et al., 2007). In the case of *Clostridium perfringens*, an enzyme of the type flavine adenine dinucleotide dehydrogenase, also capable of reducing other nitrogenated aromatic compounds, was detected.

More recently, Joe et al. (2008) presented the results obtained with *Clostridium bifermentans* SL186, closely related (99.48%) to *Clostridium bifermentans* DSM 10176, isolated from dye contaminated soil, during the biodegradation of relatively high concentrations of the following dyes: Reactive Red 3B-A, Reactive Black 5 and Reactive Yellow 3G-P. The growth of bacteria in plates in which the compounds were decolorized was the evidence that served to conclude that *Clostridium bifermentans* SL186 carried out an enzymatic Azo-reduction in this experimental work. No intermediate products (aromatic amines) were detected in this report, thus, the authors suggested that in this case, the biodegradation products of the dyes tested in the presence of *Clostridium bifermentans* SL186 may be different from the ones obtained with other anaerobic bacteria capable of biodegrading Azo-dyes.

Some of the common biodegradation products of azo-dyes under anaerobic conditions are aromatic amines such as aminobenzoate, 2 and 4-aminophenols, 2,4-dihydroxyaniline and 5- aminosalicylic acid (5-ASA). Aromatic amines derived from Azo-dye reduction can be biodegraded under methanogenic conditions, and they also can be biodegraded relatively easily under aerobic conditions (Ekici et al., 2001). For these type of compounds derived from Azo-dye biodegradation, the role of *Clostridium* sp. has not been reported.

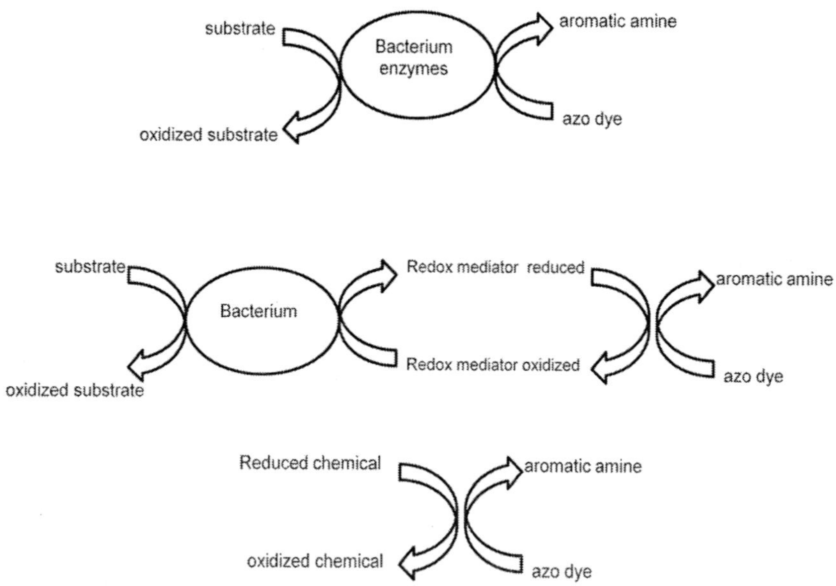

Figure 2. Mechanisms of azo-dyes biodegradation. Bacterial direct transformation (enzymatic). Biological, redox mediated and abiotic transformation.

Other strains of the fermentative *Clostridium* have been identified in consortia within engineered systems like anaerobic reactors treating wastewaters containing dyes. Often, the works intend to find the optimal operation conditions in the systems, and the choice of inocula can depend on the availability of samples of sludges, soils or sediments adapted to the pollutant. Kehlifi et al. (2009) worked on the identification of the microbial community developed in their up-flow anaerobic fixed bed reactor treating a water stream containing indigo dye, and they identified the following bacteria:

Clostridium peptidivorans, *Clostridium bifermentans*, *Clostridium pascui*, *Clostridium ganghwense* and the also fermentative bacteria *Sedimentibacter hydroxybenzoicus* along with the sulfate reducers *Desulfovibrio desulfuricans* and *Desulfovibrio vulgaris*. All these bacteria were coexisting in the reactor, yielding up to 90% removal of the dye. The inoculum of the reactor was a mixture of "non-defined" cultures including anaerobic sludge from an anaerobic digester for sulfate containing wastewater, sludge from an anaerobic reactor treating industrial wastewater and microflora from cow rumen (Kehlifi et al., 2009).

Figure 3 shows some *Clostridium* sp. involved in Azo-dyes reduction.

Clostridium sp.	Azo dye	Reference
Clostridium perfringens	Amaranth Bromo phenol blue	Kim et al. (2002) Forgacs et al. (2004) Pandey et al. (2007)
Clostridium bifermentans SL186	Reactive red 3B-A Reactive black 5 Reactive yellow 3G-P	Joe et al. (2008)
Clostridium bifermentans *Clostridium pascui* *Clostridium ganghwense* *Clostridium peptidivorans*	Indigo dye. Up-flow anaerobic fixed bed reactor	Kehlifi et al. (2009)

Figure 3. Some *Clostridium* sp. involved in azo-dyes biodegradation.

Clostridium Sp. and Biodegradation of Explosives and Their Intermediate Biodegradation Products

Contamination of extensive areas of soil and water streams including groundwater by explosives and their related compounds is a serious problem that will maintain busy, for many years to come, to a great part of the scientific community in the world. Explosives are a group of poly-nitrogenated xeno-biotic compounds of known significant negative impact on the environment all over the world.

According to one of the most recent and comprehensive reviews on biodegradation of 2,4,6-trinitrotoluene (TNT), at least in the United States, the manufacturing of TNT, by large the most used explosive, and widespread pollutant belonging to the group of the poly-nitrogenated aromatic compounds, was reinitiated in 2005, as a response of the military demands in that country (Stenuit and Agathos, 2010). Other two explosives of great importance due to the serious contamination that they cause to soil and water are Hexahydro-1,3,5-trinitro-1,3,5-triazine (RDX) and Octahydro-1,3,5,7-tetranitro-1,3,5,7-tetrazocine (HMX).

The explosives are organic nitrogenated compounds that fall within the following classifications (Rosser et al., 2001; Bhushan et al., 2003; Bhushan et al., 2004; Strigul et al., 2006). Figure 4 shows some structures of these compounds.

a) **Nitroaromatics**: 2,4,6- trinitrotoluene (TNT) and 2,4,6-trinitrophenol (Picric acid)
b) **Nitramines**: Hexahydro-1,3,5-trinitro-1,3,5-triazine (RDX) and Octahydro-1,3,5,7-tetranitro-1,3,5,7-tetrazocine (HMX)
c) **Nitrate esters**: Pentaerythritoltetranitrate (PETN) and Glyceroltrinitrate (GTN)
d) **Future generation high performance explosive (cyclic nitramines)**: 2,4,6,8,10,12-hexanitro- 2,4,6,8,10,12- hexaazaisowurtzitane (CL-20)

Contamination of the environment by explosives including the compounds mentioned above has been widely documented by several research groups that investigate alternatives for their biodegradation as well as bioremediation techniques to be applied to the many contaminated sites in which these compounds are found as the main group of pollutants.

Importance of the Biodegradation and Bioremediation Studies of Explosives in the World

In order to provide to the unfamiliar reader with a scope of the degree of contamination in the world due to these kinds of compounds, some facts taken from the report of Stenuit and Agathos (2010) regarding the pollution with explosives will be mentioned here. Sources of contamination of TNT are those related to war such as testing and training activities, bombed battlefields, leakage of military munitions, mining and quarrying activities and post-war activities. Also, TNT manufacturing activities and former ammunition facilities must be considered. Another important fact to be considered regarding TNT contamination is landmine. Further and updated data provided by Stenuit and Agathos (2010) refer to more than 70 countries affected by severe landmine contamination. Landmines are a granted source of TNT and other explosives contamination, such as the hydrophobic nitramines RDX,

HMX, and the even less water soluble CL-20, holding the possibility of leaching to groundwater streams implying both soil and groundwater contamination (Lewis et al., 2004; Bhushan et al., 2004; Stenuit and Agathos, 2010).

According to the Environmental Protection Agency in the United States (EPA, 2003), all nitro-aromatic compounds are considered priority pollutants. This, of course, includes the explosives TNT, RDX and HMX, which exhibit great toxicity to human beings, fungi, algae and microorganisms. Hydrophobic RDX, for example, is known to be human carcinogen and highly toxic to algae, invertebrates and fish at concentrations below its solubility limit in water (McLellan, 1992, cited by Zhang and Hughes, 2003). All nitro-aromatic compounds are classified with the highest level of hazardousness and toxicity (Kulkarni and Chaudhari, 2007).

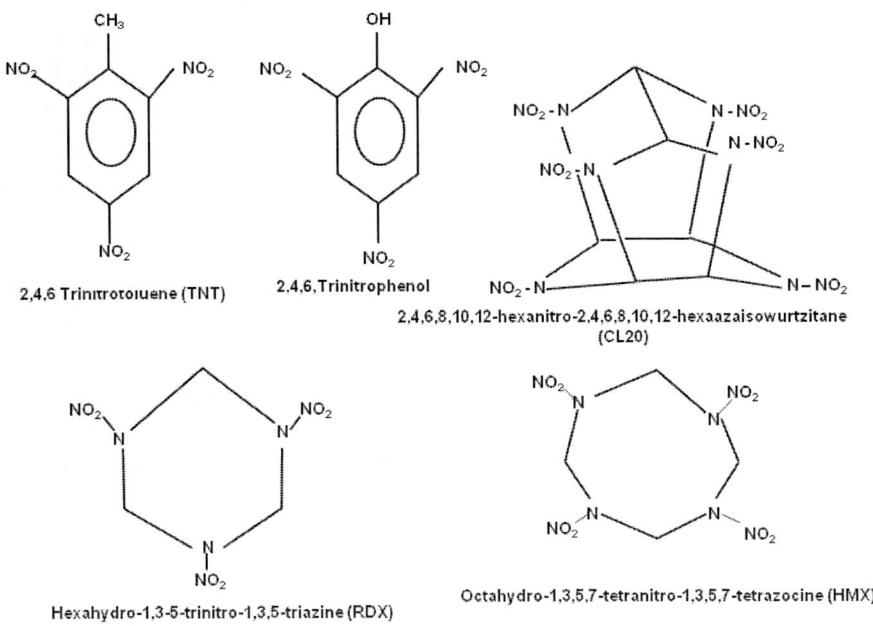

Figure 4. Chemical structures of some representative explosives.

Recalcitrance and toxicity of highly nitrogenated (or poly-nitrogenated) explosives is largely due to its chemical structure. Highly (or poly) nitrogenated aromatics present an electron-withdrawing character that makes difficult to oxidize the molecule. Therefore, unlike mono-nitrogenated aromatics and some dinitro aromatics, poly nitrogenated compounds are not susceptible to biodegradation under aerobic conditions. The higher the number of nitro groups in the molecule, the more easily reduction of the nitrogen groups will occur.

In the case of TNT, the relatively easy reduction of the nitro groups is clearly explained by Stenuit and Agathos, (2010), when they mention that the nitrogen (N) in the molecule exhibits an oxidation number of +3, thus, it can be reduced to nitroso, hydroxylamino and finally amino aromatic derivatives, which turn out to be sometimes much more toxic than TNT itself. Other reports where this TNT reduction is explained include Esteve-Núñez et al. (2001), Rosser et al. (2001), Kulkarni and Chaudhari (2007).

According to the literature, the reduction of the first nitro group is faster than the reduction of the other two nitro groups due to the thermodynamics of the reactions. The first nitro group reduction yields more free Gibbs energy derived from the redox potential (E^{o}') than the reduction of the third and last nitro group. 2,4,6 trinitrotoluene (TNT) (E^{o}' = -253 mV), 4- amino-2,6-dinitrotoluene (E^{o}'= -449 mV), 2,4-diamino-6 nitrotoluene (E^{o}'= -502 mV) (Stenuit and Agathos, 2010). In order for bacteria to use these compounds as a source of carbon and/or nitrogen, the energy yields that their uptake provides must be high enough, otherwise they may not be a substrate choice for the microorganisms and therefore will remain in the environment longer.

Fortunately, nowadays, there is a large amount of valuable information on biodegradation and biotransformation of explosives including biodegradation pathways and microorganisms involved. Within the lists of microorganisms capable of degrading explosives, several ubiquitous *Clostridium* sp. are included linked to both non-specific (cometabolic) and direct, enzymatic biodegradation of these compounds.

Under anaerobic conditions, the biodegradation of nitro groups can be enzymatic or cometabolic (involving non-specific enzymes), and it initiates with the reduction of the nitro group catalyzed by nitroreductases. There are two types of nitroreductases (Kulkarni and Chaudhari, 2007):

a) The oxygen insensitive (type I) nitroreductases, which reduce the nitro group by a pair of electrons to nitroso, hydroxylamino and amino derivatives, using as electron donor reduced pyridines (NADH and NADPH). The end products are aromatic amines and, as pointed out by several researchers, the pathway generates unstable and very reactive nitroso derivatives, which explains the fact that they are more toxic and carcinogenic than the initial nitro aromatic compound.

b) The other enzymes are the nitroreductases oxygen sensitive (type II); they reduce the nitro group to the nitroso anion due to one electron reduction. The nitroreductases classified as type II are expressed in bacteria such as *Clostridium* and *E. coli*, plants and animals.

Biodegradation of 2,4,6- Trinitrotoluene (TNT) and *Clostridium* Sp.

Biodegradation of TNT is carried out by microorganisms mainly as a result of cometabolism (Stenuit and Agathos, 2010). TNT may serve as a source of energy in nitrorespiration and/or nitrogen source, but still a carbon source or primary substrate is required to promote its biodegradation (Stenuit and Agathos, 2010).

The two genera of bacteria that have been the most studied regarding TNT anaerobic biodegradation are *Clostridium* and *Desulfovibrio* (Esteve-Núñez et al., 2001). Until very recently, it is was not completely known if the TNT metabolism by *Clostridium* sp. was inducible or in all cases was constitutive in these bacteria for which the biodegradation of TNT seems to be a shared feature (Ederer, et al., 1997; Esteve-Núñez et al., 2001). The report of Kutty and Bennett (2005) described the oxygen-insensitive nitroreductase expressed by *Clostridium acetobutylicum* ATCC 824 for the transformation of TNT showing that in the case of this strain ,the enzyme was inducible.

It has been suggested that in the case of these fermentative bacteria, the reduction of the nitro group providing a free Gibbs energy ($\Delta G°' = -94.5$ kcal/mol for the formation of 4-A-2,6- DNT) supports energetically the fermentation by reoxidation of reduced electron carriers formed during the process, thus maintaining the energy supply via substrate-level phosphorylation (Shelley et al., 1996; Ahmad and Hughes, 2000; Stenuit and Agathos, 2010).

To the best of our knowledge, there are three *Clostridium* strains for which the biodegradation pathways for TNT have been proposed, these are the closely related to: *Clostridium bifermentans* (Lewis et al., 1996), *Clostridium acetobutylicum* (Khan et al., 1997; Hughes et al., 1998a; Hughes et al., 1998b; Hughes et al., 1999) and *Clostridium thermoaceticum* (Huang et al., 2000), even though biotransformation of TNT to triaminotoluene (TAT) was reported to occur with *Clostridium pasterianum* in 1976 (McCormick et al., 1976)

Some of the biodegradation products reported for the cometabolic biodegradation of TNT by *Clostridium* sp. are TAT and aromatic nonnitrogenated compounds such as methylphloroglucinol and *p*-cresol along with phenolic products derived from TAT hydrolysis (Lewis et al., 1996). It has been proposed that the *p*-cresol identified in the TNT biodegradation is actually a byproduct of the biodegradation of aromatic amino acids such as tryptophan rather than a direct byproduct from TNT biodegradation (Shen et al., 2000; Esteve-Núñez et al., 2001). The intermediates identified in the TNT biodegradation pathway of TNT in the presence of cell extracts of *Clostridium* were: 2-amino-4-hydroxylamino-5-hydroxyl-6-nitrotoluene, a compound probably formed thorough Bamberger rearrangement of dihydroxy-dinitrotoluene (Hughes et al., 1998a; Hughes et al., 1998b).

Also, it has been suggested that the initial step for TNT anaerobic biodegradation is carried out enzymatically by the carbon monoxide dehydrogenase, an enzyme that produces the following compounds: 2-hydroxylamine-4,6-dinitrotoluene, 4-hydroxylamine-2,6-dinitrotoluene, and 2,4-dihydroxylamine-6-nitrotoluene (Huang et al., 2000). Then, the 2,4-dihydroxylamine-6-nitrotoluene, may be converted to 2-amino-4-hydroxyl-amino-5-hydroxyl-6-nitrotoluene and hydroxylamine-4-amino-5-hydroxyl-6-nitrotoluene via Bamberger rearrangement.

TNT biodegradation has been greatly studied; it seems that the main issue is not about the microorganisms capable of biodegrading it but about the application of the microbial processes in bioremediation sites contaminated with this explosive. So far, it may seem that some of the microorganisms playing main roles in the anaerobic biodegradation of TNT are explained in terms of the metabolic pathways that take in this biodegradation process. Research results suggest bioaugmentation as a promising technique for TNT bioremediation. In their comprehensive report on TNT, Stenuit and Agathos (2010) mentioned a recent work on a bacterium presumably capable of biodegrading TNT as a source of carbon and nitrogen, which shows that, as great as it seems, the knowledge on microbial capabilities may be still limited. Therefore, it is possible that other microorganisms may exhibit that feature. and it would not be surprising that in a near future. some spore-forming *Clostridium* sp. may be sharing with other bacteria some "high-tech" applications in the field.

Biodegradation of Cyclic Nitramines and *Clostridium* Sp.

On the other hand, cyclic nitramines such as RDX, HMX and 2,4,6,8,10,12-hexanitro- 2,4,6,8,10,12- hexaazaisowurtzitane (CL-20), this last one cited as a high-performance explosive of future generation (Bhushan et al. 2004; Strigul et al., 2006), posses a very low solubility in water (40, 6.6 and 3.6 mg/L respectively). Unlike TNT, which is an aromatic compound, these compounds do not present the electronic stability that TNT has; therefore, they are more susceptible to be degraded after an initial attack, either chemical or biological. The main challenge presented for the biodegradation of these compounds is the availability, since they tend to adsorb into the soils or remain inside the soil pores (Bhushan et al., 2004).

Although anaerobic biodegradation for RDX and HMX has been largely proved under different conditions including their biodegradation in soils (McCormick et al., 1981; Hawari et al., 2000; Crocker et al., 2006), in the case of CL-20, several studies have been carried out in order to identify if its biodegradation is soil is better under aerobic or anaerobic conditions (Trott et al., 2003; Strigul et al., 2002; Strigul et al., 2003). After a study carried out with different soils and different incubation conditions, it was concluded that CL20 is better biodegraded under anaerobic conditions and that the biodegradation is accelerated with addition of organic substrates, although it was not specified if the CL20 byproducts were somehow toxic to the microbial communities during the biodegradation processes (Strigul et al., 2006). Some of the isolated microorganisms from soil that have been reported to biodegrade CL20 as a sole source of nitrogen are: *Agrobacterium* sp. strain JS71 (Trott et al., 2003), *Pseudomonas* sp. strain FA1 (Bhushan et al., 2003), and *Clostridium* sp. EDB2 (Bhushan et al., 2004).

Interestingly enough, the first report on chemotaxis-mediated biodegradation of RDX, HMX and CL-20 states that one motile and flagellated strain of *Clostridium* isolated from marine sediments exhibited the capability of biodegrading the mentioned explosives by chemotaxis via N-denitration. The strain was identified as *Clostridium* sp. EDB2. During RDX, HMX biodegradation by *Clostridium* sp. EDB2 the release of NO_2^-, N_2O, HCHO, HCOOH was detected, whereas for CL-20 biodegradation by the same *Clostridium* strain, the release of NO_2^-, N_2O, HCOOH was detected. Furthermore, the research in this case demonstrated that NO_2^- elicited chemotaxis response in the cells. The rates of biodegradation obtained with the resting cells were: 1.8 ± 0.2, 1.1 ± 0.1 and 2.6 ± 0.2 n mol h^{-1} mg wet biomass for RDX, HMX and CL-20, respectively. When the cultures were amended with NADH, the biodegradation rates increased as follows: 4.5 ± 0.3, 2.5 ± 0.3 and 7.2 ± 0.6 n mol h^{-1} mg wet biomass (Bhushan et al., 2004).

Successful biodegradation of HMX to N_2O and HCHO has also been reported to occur in the presence of several other Clostridia. In a relatively recent report, Zhao et al. (2004) described the biodegradation of 36 μM of HMX by *Clostridium bifermentans* HAW-1 to N_2O cultured in two basal medium. Rate of HMX removal using yeast extract peptone-glucose medium (YPG) was 80 n mol h^{-1} mg dry biomass and 190 n mol h^{-1} mg dry biomass using reinforced Clostridial medium (RCM). Other strains of fermentative hydrogen producers *Clostridium* were also used in their work, yielding reasonable removal rates of HMX. The other *Clostridium* strains reported in that work were *Clostridium saccharolyticum* HAW-E3, *Clostridium celerecrescens* HAW-G4 and *Clostridium butyricum* HAW-HC1. In the case of these three strains of *Clostridium*, the end products of HMX biodegradation identified were HCHO and N_2O. HMX removal rate observed for *Clostridium saccharolyticum* HAW-E3 in YPG medium was of 150 n mol h^{-1} mg dry biomass and 50 n mol h^{-1} mg dry biomass using RCM. For *Clostridium celerecrescens* HAW-G4, the HMX removal rates were: 110 and 30 n mol h^{-1} mg dry biomass for YPG and RCM basal medium, respectively. Finally, for *Clostridium butyricum* HAW-HC1, the HMX removal rates were: 40 and 10 n mol h^{-1} mg dry biomass for YPG and RCM medium, respectively. *Clostridium bifermentans* HAW-1 yielded the highest HMX removal rate in RCM. The enzymes presumably involved in the biodegradation of HMX in the work of Zhao et al. (2004) were one for denitration (hydrogenase) and another for the reduction of the nitroso derivatives (nitroreductase).

A work specifically focused on RDX and the phylogenetic diversity during RDX biodegradation (Zhao et al., 2003) presented results on RDX biodegradation to CH_3OH, HCHO and N_2O by an isolate closely related to *Clostridium saccharolyticum* (isolate HAW-E3) at a removal rate of 6.2 μ mol h^{-1} mg dry biomass in YPG medium, whereas for an isolate closely related to *Clostridium celerecrescens* (isolate HAW-G4), the removal rate of RDX was 5.2 μ mol h^{-1} mg dry biomass in the same medium. In that work, it was demonstrated that these fermentative bacteria were also capable of utilizing RDX as a nitrogen source. Another bacteria identified in the RDX enrichment prepared in this work were closely related to *Clostridium bifermentans* and *Clostridium butyricum*.

Once confirmed that the cyclic nitramines RDX and HMX can be biodegraded by *Clostridium* sp. linked to bacterial growth, the enhancement of the biodegradation rates of these compounds by *Clostridium* sp. EDB2 has been proved to occur in the presence of natural occurring redox mediators such as iron (III), humic acids (HA) and quinones such as anthraquinone 2-6-disulfonate (AQDS), compounds proved to enhance biodegradation rates of compounds such Azo-dyes and chlorinated solvents (Van der Zee et al., 2001;

Field and Brady, 2003; Guerrero-Barajas and Field, 2005), even though in the case of the mentioned compounds the microbial source has been mostly methanogenic anaerobic granular sludge or consortia highly enriched in *Dehalococcoides* sp., *Desulfovibrio* and *Clostridium* sp. (Hu et al., 2011).

The role of redox mediators was evaluated in biodegradation of RDX and HMX by *Clostridium* sp. EDB2 as reported by Bhushan et al. (2006). In this work, the abiotic and biological degradation of these cyclic nitramines was tested in the presence of iron (III), HA and AQDS. The work demonstrated that RDX and HMX biodegradation occurs by two (simultaneous) different pathways: initial N-denitration and initial reduction of the nitro group to nitroso, which results in the formation of N_2O and HCHO. The list of intermediates also included nitrite, methylendinitramine, ammonium and formic acid. Final mineralization of the explosives was confirmed through $^{14}CO_2$ formation. The biological degradation of RDX was slightly faster in the presence of HA than in the presence of iron (III) and AQDS. Iron (III) and AQDS enhanced the biodegradation of the RDX over time against the test in which no redox mediator was added. Mineralization of RDX was higher in the presence of the redox mediators versus the tests in which no redox mediator was used.

For HMX, the addition of either iron (III), HA or AQDS enhanced the biodegradation and mineralization of the compound to a different extents, without any redox mediators. $^{14}CO_2$ formation was less than the obtained in the presence of them. Abiotic tests in which the transformation of the both explosives was attempted by adding either one of the redox mediators did not yield transformation results (Bhushan et al., 2006).

In our opinion, the conclusion of the work of Bhushan et al. (2006) on RDX and HMX redox mediated biodegradation is important because provides information on the metabolic behavior of *Clostridium* sp. EDB2, that may be also exhibited in other bacteria if further experimental work is conducted on pure strains, and second, it provides an insight on how useful a bacterium could be if the appropriate conditions in the site could be provided or if the information is used in designing engineered systems (i.e., medium- or large-scale bioreactors).

The conclusions issued by Bhushan et al. (2006) can be summarized graphically, as it is shown in Figure 5, a figure reproduced from their work in agreement with the authors. The interesting facts derived from this work can be briefly stated as follows (Bhushan et al., 2006):

a) *Clostridium* sp. EDB2 can biodegrade directly RDX and HMX.

b) *Clostridium* sp. EDB2 can reduce iron (III) to reactive iron minerals, which in turn may degrade RDX and HMX.
c) *Clostridium* sp. EDB2 reduces the redox mediators (or electron shuttles) HA and AQDS, which in turn can transfer electrons to the RDX and HMX leading to their biodegradation.

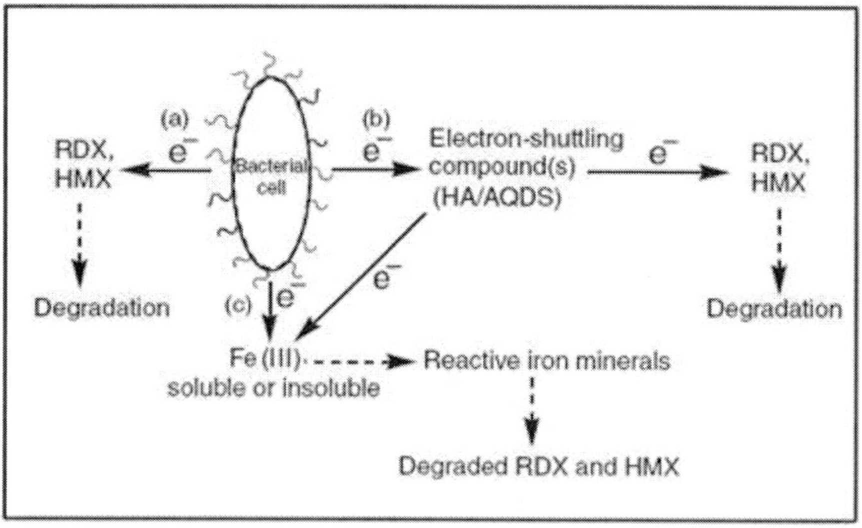

Figure 5. Schematic representation of strategies used by *Clostridium* EDB2 to degrade RDX and HMX via different routes. (a) cyclic nitramines degraded by direct interaction with the strain EDB2. (b) electron shuttling compounds (HA/AQDS), reduced by strain EDB2, abiotically degraded cyclic nitramines. (c) iron minerals, reduced by strain EDB2, abiotically degraded cyclic nitramines. Solid arrows indicate direction of flow of electrons. This figure has been reproduced from Bhushan et al. (2006), *Journal of Applied Microbiology*, 100:555-563 with permission from John Wiley & Sons, under license number: 2696760919267.

Biodegradation of Mixtures of TNT, RDX and HMX and *Clostridium* Sp. Involved

Recent studies on biodegradation of mixtures of TNT, RDX and HMX combined with the identification of the microbial communities and the possible variations that they may suffer during the biodegradation of the explosives show that *Clostridium nitrophenolicum* and *Clostridium kluyveri* DSM-555 are the two bacteria most involved in the biodegradation process (Moshe et al., 2009).

In their study, Moshe et al. (2009) worked with the three contaminants and found that TNT inhibits RDX and HMX biodegradation, and that RDX and HMX tend to biodegrade faster that TNT. *Clostridium nitrophenolicum* was present in all treatments, and it seems that it can degrade TNT and resist its toxicity. *Clostridium kluyveri* DSM-555 was present in all treatments except for the ones containing TNT, suggesting that it is capable of biodegrading RDX and HMX only. One of the important highlights of this work is the sequential biodegradation of TNT rather than simultaneous. Once TNT was biodegraded in the mixture, they were able to confirm the further biodegradation of RDX and HMX.

Since the TNT and other new explosives production in the world certainly will continue, the recent trend is to produce formulations of "biodegradable" explosives including spores of bacteria in the preparation due to the stability that they may offer.

Clostridium sp.	Explosive, intermediates or biodegradation products	Reference
Clostridium acetobutylicum	2,4,6 trinitrotoluene (TNT) → triaminotoluene (TAT)	Khan et al., 1997 Ederer et al., 1997 Esteve-Núñez et al., 2001
Clostridium bifermentans KMR-1	TNT → amino toluenes TNT biodegradation potential use of the spores of the Clostridium bifermentans KMR-1 in formulations of "biodegradable" explosives.	Stenuit and Agathos, 2010
Clostridium nitrophenolicum	TNT → tetranitroazoxytoluene (TNAzoxyT) → tetranitrohydraazotoluene → further reduction to diamino nitro toluene. RDX → hexahydro-1-nitroso-3,5-dinitro-1,3,5-triazine (MNX) HMX → octahydro-1-nitroso-3,5,7-tetranitro-1,3,5,7-tetrazocine	Moshe et al., 2009
Clostridium kluyveri DSM 555	RDX → hexahydro-1-nitroso-3,5-dinitro-1,3,5-triazine (MNX) HMX → octahydro-1-nitroso-3,5,7-tetranitro-1,3,5,7-tetrazocine.	Moshe et al., 2009
Clostridium bifermentans	TNT → RDX. TNT → TAT to 2,4,6 trihydroxytoluene	Kulkarni and Chaudhari, 2007 Lewis et al., 1996 Lewis et al., 2004
Clostridium bifermentans CYS-1	TNT → aliphatic polar compounds via 4ADNT and 2,4 DANT.	Shim et al., 1995
Clostridium bifermentans LJP-1	TNT → TAT and phenolic compounds TNT → TAT	Lewis et al., 1997 McCormick et al., 1976
Clostridium pasterianum *Clostridium sordelii*	TNT → TAT	Ederer et al., 1997 Esteve-Núñez et al., 2001

Figure 6. (Continued).

Clostridium sp.	Explosive, intermediates or biodegradation products	Reference
Clostridium acetobutylicum ATCC824	RDX → monitroso, monohydroxylamino, mononitrosomonohydroxylamino, monoamino, diamino and triamino compounds	Zhang and Hughes, 2003
Clostridium thermoaceticum	TNT → 2-hydroxylamine-4,6-dinitrotoluene, 4-hydroxylamine-2,6-dinitrotoluene and 2,4-dihydroxylamine-6-nitrotoluene	Huang et al., 2000 Esteve-Núñez et al., 2001
Clostridium nitrophenolicum	p-nitrophenol, it may be a potential degrader of explosives	Suresh et al., 2007
Clostridium sp. EDB2	RDX, HMX → NO_2^-, N_2O, HCHO, HCOOH	Bhushan et al., 2004 Bhushan et al., 2006
Clostridium sp. EDB2	CL-20 → NO_2^-, N_2O,HCOOH	Bhushan et al., 2004 Lewis et al., 2004
Clostridium bifermentans HAW-1	HMX → N_2O	Zhao et al., 2004
Clostridium saccharolyticum HAWE3	HMX → HCHO and N_2O	Zhao et al., 2004
Clostridium celerecrescens HAW-G4	HMX → HCHO and N_2O	Zhao et al., 2004
Clostridium butyricum HAW-HC1	HMX → HCHO and N_2O	Zhao et al., 2004
Clostridium saccharolyticum HAW-E3	RDX → CH_3OH, HCHO and N_2O	Zhao et al., 2003
Clostridium celerecrescens HAW-G4	RDX → CH_3OH, HCHO and N_2O	Zhao et al., 2003
		Zhang et al., 2003
Clostridium acetobutylicum	TNT → TAT	Esteve-Núñez et al., 2001
Clostridium bifermentans CYS-1	TNT → aliphatic polar compounds	
Clostridium bifermentans LJP-1	TNT → TAT and phenolic compounds	Esteve-Núñez et al., 2001
Clostridium pasterianum	TNT →TAT	
Clostridium sordelii	TNT →TAT	
Clostridium sp.	Bamberger rearrangement of dihydroxylaminodinitrotoluene.	Esteve-Núñez et al., 2001

Figure 6. Some of the most studied *Clostridium* sp. featuring biodegradation of explosives.

Spores production of *Clostridium bifermentans* KMR-1 was proposed earlier for bioremediation of nitroaromatics in1997 (Sembries and Crawford, 1997).

As previously described in the review prepared by Stenuit and Agathos (2010), a mixture of explosives such as TNT and ammonium nitrate with spores of bacteria (*Bacillus* sp.) protected with some kind of coating such as powdered wood is expected to function as "bio-explosive," as some reports have suggested (Nyanhongo et al., 2009; Dario et al., 2010). Once water is in contact with the explosive materials, the bacteria start to act. This has been tried with *Bacillus* sp. SF, and this type of formulation may be attempted also with spores of *Clostridium bifermentans* KMR-1.

Due to the complexity that contamination of explosives present, nowadays, the main trends on biodegradation and bioremediation of these compounds rely on phytoremediation and bioaugmentation. Bioaugmentation may still require further knowledge on the role and mechanisms that the microorganisms such as members of Clostridia family carry out in the biodegradation processes of these pollutants.

Finally, Figure 6 summarizes the *Clostridium* bacteria that have been involved in explosives biodegradation.

Clostridium Sp. Related to the Biodegradation of Chlorinated Compounds

Damage to the environment including damage to human health is also caused by the great variety of chlorinated compounds. These xenobiotics are very harmful, regardless of the apparently simple structure that they may have starting with relatively small molecules such as the halomethanes: chloromethane, dichloromethane and the highly toxic chloroform or carbon tetrachloride. The chlorinated aliphatics such as vinyl chloride (VC) or trichloroethylene (TCE), both of them widely known as carcinogenic compounds, and TCE are included in the list of priority pollutants of the Environmental Protection Agency (EPA) in the United States. Chlorinated solvents and the related chlorinated aliphatic compounds have received much attention in biodegradation and bioremediation since groundwater streams around the world are frequently contaminated by them in relatively large concentrations, so much so that nowadays, research is still focused on the aerobic and anaerobic biodegradation of these compounds and the technologies to apply in the several contaminated sites. As in other cases, the chemistry of these compounds determines if the mechanisms of biodegradation can be aerobic or anaerobic, and the scientific community has given a classification of higher chlorinated or lower chlorinated compounds referring to the chlorinated methanes and chlorinated solvents. Lower chlorinated halomethanes and solvents (aliphatic chlorinated ethanes and ethenes) are susceptible to aerobic biodegradation, whereas higher chlorinated halomethanes and solvents tend to be readily biodegraded under anaerobic conditions.

A very useful classification of these chlorinated solvents and related halogenated aliphatics is the given by Field and Sierra-Alvarez (2004), which presents the following categories for these compounds: chloromethanes, chloroethanes, choroethenes, chlorofuorocarbons, chloroacetic acids, chloropropanes, chloropropenes and epichlorohydrin and chlorobutadienes. A great survey on the production of these compounds in the main industrialized countries prepared at that time by these authors shows that several tons/year may contribute to the release of them to the environment through inefficient disposal practices in addition to the natural sources that are also taken into account in the production numbers. In this vast review (Field and Sierra-Alvarez, 2004), the biodegradation under aerobic and anaerobic conditions for these compounds is explained, and it also includes the type of microorganisms that are involved in each case.

Without a doubt, toxicity of several of the compounds falling into the categories mentioned above can compete with the adverse effects that azo-dyes, explosives, and even metals may eventually cause to the environment.

In addition to the chlorinated aliphatics, other chlorinated compounds that imply a serious contamination problem are polychlorinated biphenyls (PCB's), dioxins, and pesticides, none of them less harmful than the other.

Taking into account that the present chapter is a survey on the role of *Clostridium* sp. in the biodegradation of chlorinated compounds, only the most important facts on the biodegradation of the compounds in which *Clostridium* are involved will be mentioned. The reader may want to refer to the vast amount of information that can be found on specific biodegradation for each classification of chlorinated or halogenated compounds to gain a perspective on the special microbiological complexity that they involve.

Under anaerobic conditions, highly chlorinated compounds may be biodegraded by cometabolism (non-specific enzymatic) or dehalorespiration (known also as halorespiration, enzymatic); in both cases, the involvement of *Clostridium* sp. has been demonstrated for some of the highly halogenated compounds.

Clostridium sp. have been related to the cometabolic biodegradation of 1,1,1- trichloroethane (1,1,1-TCA) cometabolically, in the same fashion, *Clostridium pasteurianum* has been reported to biodegrade chlorofluorocarbon-12 (CFC-12) even though at much slower rate than methanogenic cultures (at an activity of 0.005 mg/g cell dry weight, D. Lovley et al., 1992; Field and Sierra-Alvarez, 2004).

Dehalorespiration, which is the capability of reducing the halogenated compound as terminal electron acceptor yielding energy for growth, has been exhibited by *Clostridium bifermentans* DPH-1 in the presence of tetrachloroethylene (PCE) producing *cis*- 1,2-dichloroethene (cDCE) as the final product using yeast extract and glucose as electron donors (Chang et al., 2000). The bacterium was isolated from a site apparently contaminated with this type of compound. Chang et al. (2000) also reported the following compounds that *Clostridium bifermentans* DPH-1 was capable of biodegrading: trichloroethylene (TCE), cDCE, *trans* 1,2-dichloroethene (tDCE), vinyl chloride (VC), 1,2 dichloroethane, 1,2 dichloroethane, dichloromethane, 1,3 dichloropropene, 1,2 dichloropropene, 1,1,2 trichloroethane and 1,1,1 trichloroethane (Chang et al., 2000). The enzyme responsible for the biodegradation of PCE to TCE and finally to c-DCE expressed by *Clostridium bifermentans* DPH-1 has been characterized showing a rate of PCE reduction of 73 nmol/mg protein. The maximum activity of this enzyme was at 35°C and pH 7.5. The approximate molecular weight is 70 KDa, and it is oxygen sensitive; it was demonstrated that its activity is independent of metal ions (Okeke et al., 2001).

More recently, Hata et al. (2004) isolated *Clostridium* sp. strain DC1 closely related to *Clostridium saccharobutylicum* (98.9% similitude) from landfill samples and demonstrated its capability of biodegrading *cis* -1,2-dichloroethene (cDCE) and vinyl chloride not linked to growth (dehalo-respiration), suggesting a mechanism of oxidation of the compound instead of reduction, though it was not determined to which reduction reaction this oxidation was coupled.

On the other hand, chlorinated compounds owing a more complex structure like the known carcinogen pesticide dieldrin (1,2,3,4,10,10-hexachloro-6,7-epoxy-1,4,4a,5,6,7,8,8a-octahydro-exo-1,4-endo-5,8-dimethanonaphthalene), which tends to accumulate in soils and sediments, is susceptible of biodegradation under anaerobic conditions as reported by Chiu et al. (2005), who investigated its biodegradation (up to 100 μg/mL) using contaminated sediments taken from a river in Taiwan. Dieldrin was biodegraded to Aldrin (a molecule simpler than dieldrin) presumably by epoxide reduction, and the presence of bacteria belonging to *Clostridium* genus was confirmed. These bacteria included *Clostridium acidaminophilum* and *Clostridium* sp. XB90, along with other uncultured bacteria, the genus *Clostridium* being the most predominant, and it is suggested that these microorganisms (*Clostridium* sp.) were responsible for the epoxide reduction that led to Dieldrin biodegradation (Chiu et al., 2005)..

Insecticides used in agriculture such as γ-hexachlorocyclohexane (γ -HCH or lindane) and some other isomers of it like α and β hexachlorocyclohexanes have been reported to be biodegraded by *Clostridium rectum* in earlier years (Ohisa et al., 1980).

Polychlorinated biphenyls (PCB´s or Aroclors, which is their trade name) are an important group of organic chlorinated pollutants for which research efforts on biodegradation and bioremediation will continue in the near and far future. Toxicity of these compounds is widely known and largely due to their lipophylic characteristics, which make them susceptible to bioaccumulation and biomagnifications along with the damage that it involves to the biota and human being. There are some remarkable relatively recent reviews on the biochemistry of certain microorganisms in the presence of these compounds addressing the mechanisms and possibilities of biodegradation of them in the environment either aerobic or anaerobically (Abraham et al., 2002; Furukawa and Fujihara, 2008; Field and Sierra-Alvarez, 2008b). Despite the several possibilities of aerobic and anaerobic biodegradation of PCB´s, *Clostridium* sp. are not precisely featuring the biological transformations of these compounds. Some reports mention the presence of Clostridia in the microbial communities involved in certain PCB´s biodegradation, but to the best of our knowledge, there is no clear evidence that any *Clostridium* alone is responsible for significant biotransformation in the presence of these compounds even though consortia enriched in *Clostridium* sp. may be.

Clostridium sp.	Organic pollutant	Reference
Clostridium acidaminophilum	Dieldrin	Chuan-Chiu et al., 2005
Clostridium beijerinckii	Para and meta- polychlorinated biphenyls (PCB`s)	Hou and Dutta, 2000
Clostridium hydroxybenzoicum		
Clostridium cellobioparum		
Clostridium intestinalis		
Clostridium cellulosi		
Clostridium botulinum		
Clostridium proteolyticum		
Clostridium bifermentans DPH-1	Tetrachloroethylene (PCE) to cis-dichloroethylenene (cDCE) via dehalorespiration. Trichloroethylene (TCE)	Chang et al., 2000 Okeke et al., 2001
Clostridium sp. strain DC-1	cis-1,2-dichloroethylene (cDCE) and vinyl chloride (VC) via anaerobic oxidation.	Hata et al., 2004

Figure 7. *Clostridium* sp. involved in biodegradation of some chlorinated compounds.

The report based on experimental work carried out by Hou and Dutta (2000) lists nine *Clostridium* sp. identified during the biodegradation of *meta* and *para*- PCB´s using samples of PCB´s contaminated sediments. The *Clostridium* sp. reported in this work were: *Clostridium beijerinckii*, *Clostridium hydroxybenzoicum* (strain 1), *Clostridium cellobioparum*, *Clostridium intestinalis*, *Clostridium cellulosi*, *Clostridium botulinum*, *Clostridium proteolyticum*, *Clostridium hydroxybenzoicum* (strain 2) and *Clostridium thermolacticum*. This report suggests that in long-term contaminated sites, it may be possible to identify more *Clostridium* sp. active on PCB´s biodegradation.

Bedard et al. (2006) presented a work on a sediment-free culture enriched in microorganisms capable of biodegrading a highly chlorinated commercial PCB´s mixture. Among the microorganisms identified in the active culture, these investigators found some bacteria belonging to four Clostridial families, being them closely related to *Sedimentibacter* (100% similitude), *Lachnospiracae* (94% similitude), *Clostridiacae* (64% similitude) and *Syntrophomonadaceae* (96% similitude).

Finally, to the best of our knowledge, so far, *Clostridium* sp. have not been linked to the biodegradation of another major group of chlorinated compounds and known toxic pollutants for which the reader may find literature surveys on their biological degradation: dioxins (Field and Sierra Alvarez, 2008a).

Some of the *Clostridium* sp. involved in chlorinated compounds biodegradation are listed in Figure 7.

Biotransformation of Metals Involving *Clostridium* Sp.

It is well known that several bacteria and archaea play an important role on the biogeochemical cycling of metals and metalloids and that they are also capable of reducing radionuclides, a fact that turns out to be fortunate because under certain conditions, many of them are undesirable.

The toxic metals for which there is no known biological function are: Ag, Cg, Sn, Au, Hg, Tl, Pb, Al and the metalloids Ge, As, Sb, and Se. Radionuclides such as Uranium are also toxic. Speciation of a metal is by large the main factor that causes toxicity, because it changes the metal availability in the environment.

In this chapter, we will refer to two main processes in which Clostridia are known to be involved: a) the reduction of some metals including the radionuclides Uranium and Plutonium during the decontamination of groundwater and sediments, and b) acid mine drainage (AMD). At this point, it is convenient to mention that regarding metals reduction and AMD, *Clostridium* sp. are most of the time working together with sulfate-reducing bacteria (SRB) and that it is difficult to completely separate SRB from the topic.

AMD is, taken as a whole, a process that leads to high concentrations of metals and sulfate at very low pHs as a result of mining activities. SRB such as *Desulfovibrio* sp. and the fermentative *Clostridium* sp. are also closely related to the microbial sulfate reduction process that often is used in the bioremediation of the streams of water derived of AMD, suggesting that the success of the metals precipitation by sulfate reduction is the result of the combined metabolism of the microorganisms belonging to both genera. Given the fact that this chapter is focused on the role that Clostridia may play in such transformations, the information presented here will only deal with the metals for which Clostridia have shown influence, even though the *Clostridium* involved may be related to SRB activity. The chapter will also mention some cases in which the microbial communities identified in AMD bioremediation processes include Clostridia in the phylogenetic schemes, even though at this point it cannot be completely neglected the role that SRB and *Clostridium* sp. play together. Other genera of bacteria that are capable of metals bio-transformation and have also been associated with Clostridia in the process are *Shewanella*, *Geobacter* and *Desulfotomaculum* (Meriah Arias and Tebo, 2003; Lloyd, 2003; Martins et al., 2010a).

On the other hand, groundwater, sediments and soil contamination by Uranium is a problem derived from mining and milling activities and manufacturing of nuclear weapons, and it is a problem detected in the United States and Europe. Uranium contamination is undesirable due mainly to its known nephrotoxicity more than its radioactivity (Madden et al., 2007; Akob et al., 2008; Boonchayaanant et al., 2009; Martins et. al., 2010a). Regarding Chromium (as Cr (VI)), this is far more toxic to the microorganisms than U(VI), and its toxicity to human beings and the environment in general is also higher, since it is considered mutagenic, carcinogenic and teratogenic. Contamination by Cr (VI) is also extended to soils and groundwater streams in the US and Europe as the result of mining processes (Martins et al., 2010a; Stewart et al., 2010). Reduction of U (VI) and Cr (VI) reduces their mobility and therefore the possibility of groundwater contamination by leaching.

Several research groups around the world have conducted lab and field scale studies on biotransformation of metals and radionuclides as well as AMD bioremediation reporting the microbial communities involved. Consortia and in some cases pure cultures of certain Clostridia have been examined in regards of this goal. Mechanisms of Uranium and Chromium reduction by some Clostridia are assumed to be enzymatic, whereas during the precipitation of metals associated to AMD like Fe, Cu and Zn, the enzymatic activity to directly reduce the metal is attributed to SRB as a metal resistance mechanism. In this case, *Clostridium* sp. are considered to maintain a helpful syntrophic relationship with the sulfate reducers, i.e., *Desulfovibrio* in order for the metal sulfides precipitation to occur (Alexandrino et al., 2011).

In the work conducted by Gao and Francis (2008), four pure cultures of *Clostridium* sp. were tested on U (VI) reduction under different pH conditions using glucose as electron donor. It was suggested that the reduction was enzymatic since it only occurred in the presence of growing or resting cells.

The authors presented the rates of U (VI) reduction as follows: *Clostridium sp* 83 $\mu M\ h^{-1}\ g^{-1}$ protein at pH 4.2, *Clostridium acetobutylicum* 12.4 $\mu M\ h^{-1}\ g^{-1}$ protein, at pH 5.8, *Clostridium pasteurianum* 7.2 $\mu M\ h^{-1}\ g^{-1}$ protein at pH 5.8 and *Clostridium sphenoides* reduced U(VI) at a rate of 31.3 $\mu M\ h^{-1}\ g^{-1}$ protein, at pH 6.5. Additionally, the experiments using nitrate at the same time as U(VI) demonstrated that U(VI) was not affected and therefore, this fact supported the assumption that in these isolates of *Clostridium*, the process was nitrate independent, which confirms what other research groups had also reported previously (Madden et al., 2007). Furthermore, the report of Francis et al. (2008) on anaerobic reductive dissolution of Pu (IV) to Pu (III) was confirmed by *Clostridium* sp. ATCC 53464 to which reduction of the following metals is also attributed: Fe (III) to Fe (II), Mn (IV) to Mn (II), Tc (VII) to Tc (IV) and U (VI) to U (IV). The experimental work of these researchers was performed with ^{242}Pu (IV)-nitrate mixture in the culture and the reduction was confirmed through selective extraction of the culture by thenoyltrifluoroacetone (TTA), indicating the presence of the reduced Pu in the soluble fraction.

Just after Gao and Francis (2008) reported that the isolates of *Clostridium* that they tested were able to enzymatically reduce U(VI), Dalla-Vecchia et al. (2010), based on the work of Gao and Francis, tested spores of *Clostridium acetobutylicum* on U(VI) reduction in order to know if this characteristic was actually exclusive of *Desulfotomaculum reducens* MI-1. The results obtained by this research group showed that the spores of *Clostridium acetobutylicum* are indeed capable of U(VI) reduction but in order for the reduction to occur, the biological reaction between the spores and the U(VI) must be carried out in the spent growth medium in which the vegetative cells were cultured. It remains unknown what kind of compound suspended in the spent medium is required for this reaction to occur, but it is assumed that it is something released by the vegetative cells. There is evidence that the reduction of U(VI) occurs directly on the spores because the reduced Uranium (U(IV)) was present on the spore surface. The work of Dalla-Vecchia et al. (2010) provides an insight on novel reduction mechanisms in bacteria and suggests that the enzyme responsible for the reduction may be located in the exosporium of the cell.

In order to address the potential problem that reoxidation of U(IV) to U(VI) may imply, Boonchayaanant et al. (2009) investigated the reoxidation of U(VI) under sulfate- and iron-reducing conditions, both conditions that are likely to occur in the environment. Their findings showed that reoxidation of the Uranium does occur, but it occurs to a less extent under iron-reducing conditions. The clone libraries prepared by these researchers indicated that approximately 74% of the bacteria involved in the iron-reducing cultures belonged to the *Clostridium* genus. Certainly, *Clostridium* sp. can also effectively reduce Fe (III) to Fe (II), but it is typical to report *Geobacter* sp. under iron-reducing conditions. In the work of Boonchayaanant et al. (2009), *Geobacter* sp. was present in a relatively low percentage (4.3%) in the sequences analyzed, and the authors explained this as probably due to the use of ethanol as electron donor.

The efforts on bioremediation of sites contaminated with Uranium have led to complementary work on U (VI) reduction under iron- and sulfate-reducing conditions in order to analyze the effect of the different electron donors on the microbial communities. Some of these experimental data have shown that Clostridia are present along with the following genera: *Geobacter*, *Tolumonas*, *Arthrobacter*, *Dechloromonas* and *Pseudomonas*, and that the composition of the microbial communities depends on the use of ethanol or glucose (Akob et al., 2008).

A strong and recent evidence that *Clostridium* sp. are active during the precipitation of sulfides of Fe, Cu and Zn and supportive to sulfate reduction process together to *Desulfovibrio* sp. is the provided by Alexandrino et al. (2011). In this recent work, concentrations of Fe (0.75 g/L), Zn (0.2 g/L) and Cu (0.080 g/L) along with 3.9 g/L of sulfate ion were included in the enrichment culture of bacteria resulting in 100% precipitation of Fe, 96% Zn and 97% of Cu in approximately one month of incubation. The results of this research group highlight the syntrophic relationship between Clostridia and

Desulfovibrio, which allowed high sulfate reduction efficiency and at the same time high resistance to the metals. The bacteria identified in this enrichment culture after 16S rRNA analysis showed that they were closely related to *Clostridium subterminale* (99% similitude), *Clostridium pascui* (98% similitude), *Clostridium mesophilum* (100% similitude), *Clostridium peptidovorans* (99% similitude) and *Desulfovibrio desulfuricans* (99-100% similitude of several clones). Another interesting fact provided by Alexandrino et al. (2011) is that the sources of microorganisms for which higher sulfate reduction was achieved were enrichments of samples taken from soils surrounding fumaroles in hydrothermal and geothermal areas, suggesting that these environments are relatively unexplored with regard to the biochemical activity of their microbial communities since the evidence showed that the performance on sulfate reduction and metals resistance was higher than the reported in other works in which different source of bacteria, usually coming from contaminated sites have been used (Martins et al., 2010b; Alexandrino et al., 2011).

After reviewing the findings on the improvement of sulfate reduction obtained by Alexandrino et al. (2011) in enrichments where Clostridia have been present, it is worth mentioning that efforts are being made to further clarify the role of *Clostridium* sp. within the Sulfur cycle. It is known that another characteristic of *Clostridium* sp. is the non-dissimilatory sulfate reduction capability, but there is still work to do regarding the capability of reducing other sulfur species such as polysulfides. Although it has not been proven that all *Clostridium* may enhance sulfate reduction in mixed cultures and therefore the precipitation of metals, it can be said that some members of Clostridia may favor the sulfur cycle by reducing sulfate and sulfur reduction intermediates such as the polysulfides, which eventually could favor the production of hydrogen sulfide and the precipitation of metals. In sight of this possibility, some findings on the role of *Clostridium* sp. in certain parts of the sulfur cycle will be presented in this section a little further; the intention will be to suggest that some more strains of *Clostridium* may actually be more active in this cycle and therefore more useful in certain applications of environmental microbiology such as precipitation of metals.

Under sulfate-reducing conditions, SRB can coexist with *Clostridium* as it has been mentioned earlier, and it may be possible that *Clostridium* enhance is some cases the yields of sulfate reduction mainly due to the capability of some *Clostridium* strains to reduce polysulfides as recently suggested by Takahashi et al. (2010) for polysulfide reduction.

The polysulfides (S_n^{-2}) are intermediate products occurring during sulfate/sulfur-reducing conditions, although chemical, abiotic reactions lead to the formation of these intermediates, it has been proven that they can be biologically reduced and that unlike sulfur, they are bioavailable to microorganisms due to their solubility.

Polysulfides yield sulfur and hydrogen sulfide according to the following reactions (Takahashi et al., 2010):

[1] $H^+ + S_x^{2-} \rightarrow HS^- + (x-1)\ S$ (precipitated), the precipitated is still more soluble than elemental sulfur.
[2] $HS^- + (x-1)\ S \leftrightarrow H^+ + S_x^{2-}$
[3] $(x-1)\ S_{x+1}^{2-} + HS^- + OH^- \leftrightarrow x\ S_x^{2-} + H_2O$

Equilibrium reactions (2) and (3) occur among polysulfides, hydrogen sulfide and elemental sulfur.

Known sulfur-reducing microorganisms belong to the genus *Desulfuromonas* and according to the literature, so far, *Wolinella succinogenes* is the only confirmed microorganism capable of reducing polysulfides. Quoting Takahashi et al. (2010): "Although elemental sulfur-reducing strains can reduce polysulfides, not all polysulfide-reducing strains can reduce elemental sulfur." Thus far, it may be clear that is interesting to unveil the role of the microorganisms, whichever they may be, during the reduction of sulfur and polysulfides as is the case of the work reported for Clostridia. In this regard, the results obtained by Takahashi et al. (2010) are very insightful. Just as in the case of Alexandrino et al. (2011), who used samples from hydrothermal and geothermal areas, Takahashi et al. (2010) prepared enriched cultures with samples from hydrothermal and hot springs areas and some sites contaminated with aliphatic chlorinated compounds such as trichloroethylene (TCE) in certain areas in Japan, to investigate the microbial community responsible for polysulfides reduction developed under sulfate-reducing conditions.

Under sulfate-reducing conditions, Takahashi et al. (2010) obtained 21 isolates from polysulfide enrichment cultures, all belonging to the genus *Clostridium,* no members of the sulfate reducer genus *Desulfovibrio*, which is frequently identified in enrichments prepared under sulfate-reducing conditions were identified among the isolates obtained under these conditions. The isolates presented a high similitude to *Clostridium subterminale* DSM 6740 (99.4-99.6% similitude), *Clostridium thiosulfatireducens* DSM 13105 (99.6% similitude), *Clostridium intestinale* DSM 6191 (96.3 % similitude),

Clostridium celerecrescens DSM 5628 (99.1-99.5 % similitude) and *Desulfotomaculum gottoideum* DSM 4024 (98.9-99.3% similitude); this last one is still considered a sulfate reducer but recently proposed to be classified as *Clostridium* due to insufficient evidence of its sulfate-reducing capability.

Clostridium sp.	Metal	Reference
Clostridium subterminale *Clostridium pascui* *Clostridium mesophilum* *Clostridium peptidovorans* In association with the sulfate reducer *Desulfovibrio desulfuricans*	Mixtures Fe, Zn and Cu and sulfate reduction for the precipitation of metals. Syntrophic association with SRB. Potential remediation of AMD.	Alexandrino et al., 2011
Clostridium pascui	U(VI) → U(IV) Cr(VI) → Cr(III)	Martins et al., 2010
Clostridium sp. (ATCC 53464)	U(VI) → U(IV)	Gao and Francis, 2008
Clostridium sp. (ATCC 53464)	Pu(IV) → Pu (III) U(VI) →U (IV) Tc (VII) → Tc (IV) Fe (III) → Fe (II) Mn (IV) → Mn (II)	Francis et al.,2008
Clostridium sphenoides ATCC 19403	U(VI) → U(IV)	Gao and Francis, 2008
Clostridium acetobutylicum ATCC 824	U(VI) → U(IV)	Gao and Francis, 2008
Clostridium pasteurianum ATCC7040	U(VI) → U(IV)	Gao and Francis, 2008
Clostridium pasteurianum	Linked to V(V) reduction.	Lloyd, 2003
Clostridium acetobutylicum spores	U(VI) → U(IV)	Vecchia et al., 2010
Clostridium beijerinckii	Cr(VI) → Cr (III)	Stewart et al., 2010
Clostridium celerecrescens		García-Balboa et al., 2010 Madden et al., 2007
Clostridium sp. strain Kas106-4		Akob et al., 2008
Clostridium beijerinckii X68180		Akob et al., 2008
Clostridium chromoreductans AY 228334		Akob et al., 2008

Figure 8. Members of the genus *Clostridium* involved in the reduction of metals.

Specifically, the isolates obtained in the work were capable of reducing polysulfides to hydrogen sulfide (H_2S) using lactate and acetate as electron donors to a different extent, according to Takahashi and coworkers (2010). *Clostridium subterminale* and *Clostridium thiosulfatireducens* were also capable of reducing elemental sulfur to hydrogen sulfide (H_2S) using lactate as electron donor. Different extents of lactate consumption and sulfur reduction were determined for the several isolates and strains used in this experimental work, and the rate of consumption of the substrate was also reported.

Sulfate reduction was carried out to a minor extent by the isolates obtained and closely related to the type strains: *Clostridium subterminale*, *Clostridium thiosulfatireducens* and *Clostridium celerecrescens* when compared to the pure strains of sulfate reducers: *Desulfovibrio vulgaris* and *Desulfotomaculum ruminis* and the pure strain of bacterium *Desulfotomaculum gottoideum*. But the mentioned isolates were capable of reducing more sulfate than the pure strains of *Clostridium subterminale*, *Clostridium thiosulfatireducens* and *Clostridium celerecrescens* that were tested for the comparison. The bacterium that reduced sulfate the least was *Clostridium thiosulfatireducens*. The experiment on sulfate reduction reported in this work shows that no FeS was precipitated in the *Clostridium* type strains cultures and that the reduction of sulfate by these strains (*Clostridium subterminale*, *Clostridium thiosulfatireducens* and *Clostridium celerecrescens*) was the least efficient (Takahashi et al., 2010).

In Figure 8, some *Clostridium* sp. involved in metals reduction are listed.

APPLICATION OF *CLOSTRIDIUM* SP. IN MICROBIAL FUEL CELLS (MFCS)

Microbial fuel cells (MFCs) are such an extensive and interesting topic nowadays that the literature available on the description, design, pure strains of microorganisms and microbial communities involved in them, as well as information on substrates used and the electricity yield in these systems, would not be able to be revised and summarized before a new finding is already reported. Currently, it is one of the most studied systems to be used as an alternate energy source as it can be confirmed by the large increase in scientific reports that several countries have issued on this topic (Pant et al., 2010). Besides the generation of energy, the substrates utilized in these devices often contain organic pollutants to be degraded, and therefore the MFC presents the additional advantage of being used to decontaminate at the same time that it produces energy (Pant et al., 2010). Being this topic a clear demonstration of how useful microorganisms can be in engineered systems designed for alternative ways of generation of electricity, a brief explanation of MFCs systems will be given, and afterwards some MFCs systems in which *Clostridium* sp. have been identified will be presented.

Also, it is worth mentioning that some reviews on MFCs that present interesting data and current facts have been prepared by Rabaey and Verstraete (2005), who revised the metabolism and the key parameters that define the performance of an MFC; Du et al. (2007) presented a survey on the microbes used in these devices, the design features, the performance and operating conditions, applications of MFCs and the possible role of MFCs in the future. On the other hand, Pant et al. (2010) presented a review focused on the several substrates used for the energy generation in these systems including some organic pollutants. Franks and Nevin (2010) prepared a review mentioning the applications and potential applications of the MFCs including the most relevant microorganisms that have been used in them and the electrical interactions between microbes and the electrodes in an MFC.

An MFC system in the simplest way consists of two chambers: an anodic chamber and a cathodic chamber separated by a proton exchange membrane (PEM), generally of Nafion (Dupont, Wilmington, Delawere) and inter-connected by a saline bridge allowing the electrons flow as is depicted in Figure 9; the substrate is oxidized to CO_2 in the anodic chamber by the microorganism or the microbial community. Protons flow occurs thorough the PEM, whereas in the cathode occurs the reduction of $O_2 \rightarrow H_2O$; thus, the generation of electricity is promoted as a result of the electrons flow from the anode to the cathode (Rabaey and Verstraete, 2005; Du et al., 2007). Therefore, the reactions carried out in the MFC would be:

1) Substrate + $H_2O \rightarrow CO_2 + H^+$ (protons) + e^- (electrons). Anode half reaction. **Microbial mediated half reaction**.
2) $O_2 + 4e^- + 4H^+ \rightarrow 2H_2O$. Cathode half reaction.

Although many microorganisms are able to thrive in an MFC and generate electricity, there are some other important aspects besides the biological component to be considered for a successful performance of an MFC. As pointed out by Rabaey and Verstraete (2005), one of them is the potential in the anode, which will determine the metabolism of the MFC. According to the potential in the anode, the metabolism in an MFC can be: high oxidative potential, medium or low oxidative potential, or fermentation, and this implies that the microorganisms that can be used in an MFC can be aerobes, facultative anaerobes or strict anaerobes. In fact, in the early nineties, *Escherichia coli* and *Saccharomyces* were used in an MFC with platinum electrodes as cited by Du et al. (2007), and according to the several results of research reported in literature, basically any microorganism can be used to generate electricity in an MFC under the appropriate operation conditions (Rabaey and Verstraete, 2005; Du et al., 2007; Pant et al., 2010). A classification of MFC based on the electron donors could be 1) photoautotrophic 2) heterotrophic 3) sediments biofuel cells (Rabaey et al., 2004).

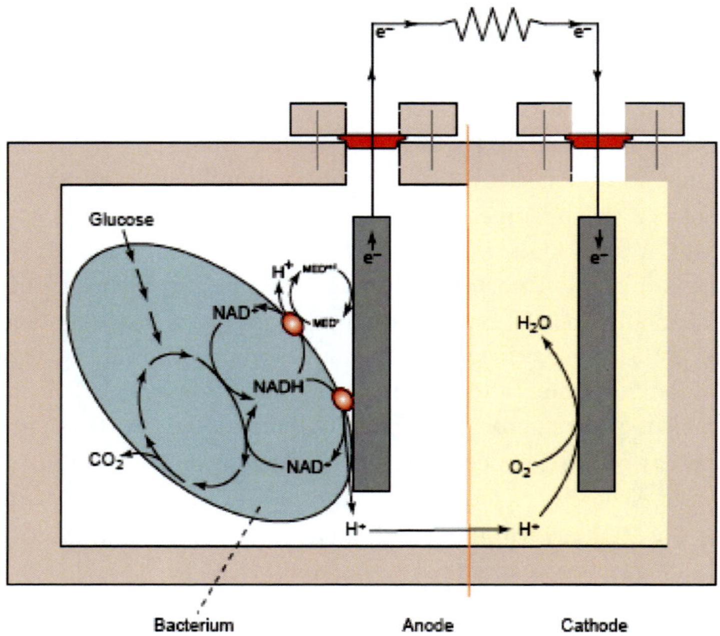

Figure 9. Scheme of a microbial fuel cell (MFC) showing the working principle. Substrate is metabolized by bacteria, which transfer the gained electrons to the anode. This can occur either directly through the membrane or via mobile redox shuttles. MED, redox mediator. Red oval, terminal electron shuttle in or on the bacterium. This figure has been reproduced from Rabaey and Verstraete (2005), *Trends in Biotechnology*, 23(6): 291-298 with permission from Elsevier, under license number: 2696771209845.

In general, when the oxidative potential in the anode is high, electron transport can be efficiently achieved by means of the biochemistry in the respiratory chain process. The energy flow (electrons and protons) is carried out by compounds such as NADH-dehydrogenase, ubiquinone, coenzyme Q and cytochromes (Kim et al., 2004; Rabaey et al., 2004). At lower anode potentials and alternative electron acceptors such as nitrate or sulfate present in the anode, the electrons sink will be the electron acceptor present, and this often indicates that bacteria are not using the anode (Rabaey and Verstraete, 2005). When the anode potential remains low, then fermentation will be likely to occur. Since transfer of electrons to the electrode is a crucial step for the generation of electricity in an MFC, an MFC can be supported by using redox mediators in order to optimize the electron transfer process when bacteria used for the process are not able to produce them. Extensive research has reported the use of several compounds that can be used as redox mediators when bacteria are not able to release electrons directly to the anode (Du et al., 2007); in these cases, the anode receives the electrons from the reduced mediator. One of the most used redox mediators in MFC is AQDS (9, 10-Anthraquinone-2,6 disulfonic acid disodium salt, (a humic model compound), which is reduced to AHQDS.

The report of Sund et al. (2007) describes the results obtained using with *Clostridium cellulolyticum* and several redox mediators for the generation of energy and fermentation products of this microorganism. They proved some compounds as redox mediators such as 9, 10-Anthraquinone-2,6 disulfonic acid disodium salt (AQDS), resazurin, methylene blue, safranine O, humic acids and cysteine. The results on current production with resazurin were higher than with the other compounds without change in the fermentation products that commonly appear in the absence of the redox mediator and suggest that resazurin is interacting with the electrode surface improving the energy flow.

Among the microorganisms that have been identified in MFCs without the addition of external redox mediators is the fermentative, Fe (III) reducing *Clostridium butyricum* EG3 that uses cytochromes as electron shuttles and that was isolated from an MFC that used for the process starch wastewater (Park et al., 2001; Rabaey and Verstraete, 2005). This fact makes *Clostridium butyricum* EG3 comparable to the electrochemically active iron reducer *Shewanella putrefaciens*, for which it is believed that the cytochromes aiding the iron (III) reduction are located in the outer membrane. Although the fermentation balance of *Clostridium butyricum* EG3 within the MFC could not be accomplished, it was demonstrated that it grows in the presence of Fe (III) and that it is electrochemically active and with similar behavior if compared to *Clostridium beijerinckii* (Park et al., 2001). The current density obtained with starch at 10 g/L with a pure culture of *Clostridium butyricum* in an MFC (graphite anode) was of 1.3 mA/cm^2, as reported in the experimental work of Niessen et al. (2004). More recently, Finch et al. (2011) reported that

Clostridium acetobutylicum can generate electricity efficiently in an MFC without the addition of external redox mediators in its metabolism of glucose.

As mentioned earlier, substrates (electron donors) for MFC include a variety of water and wastewater streams containing compounds such as azo-dyes, landfill leachates (a potential source of organic pollutants and heavy metals), brewery wastewater and starch processing wastewater besides the greatly employed acetate and glucose (Pant et al. 2010). One of the applications of MFC has been the treatment of cellulose-related compounds, as in the case of 1 g/L of Carboxymethyl cellulose (CMC), in which the culture of the MFC consisted of *Clostridium cellulolyticum* and *Geobacter sulfurreducens* (Ren et al., 2008; Pant et al., 2010) yielding a current density of 0.05 mA/cm^2 at maximum power. Biodegradation of diesel in an MFC has been reported by Morris et al. (2008 and 2009), in which work (2009), it is mentioned that *Clostridium aminobutyricum* along with an uncultured bacterium clone aaa99f09 (DQ816804) were isolated from the anode of a single chamber MFC using hydrocarbons as a substrate (electron donor).

According to research, the easier the oxidation of the substrate, the higher yield in current density (mA/cm^2). Estimations on power (voltage) can be calculated as described by Rabaey and Verstraete, (2005) from the current density data. Current density magnitude varies depending on the design of the MFC, the substrate and the microorganism used. It can present values of 0.0005 mA/cm^2, using nitrotriacetic acid (NTA) as substrate for an oligotrophic consortium, or much higher values such as 2.05 mA/cm^2 for pure cultures of *Geobacter sulfurreducens* using sodium fumarate as substrate and a different MFC design (Pant et al., 2010). So far, *Clostridium* sp, have shown to be competitive with other bacteria in the efficient generation of electricity in MFCs.

Clostridium Sp. Involved in Other Biological Processes

Olive Mill Wastewaters

Despite the several “classifications” that already exist for many organic pollutants, there are certain cases in which they are present in complex mixtures, such as it may be in industrial wastewaters. One of the cases in which the mixture of compounds in the wastewater is highly toxic and represents a serious problem is olive mill wastewater. Olive mill wastewater is a mixture of proteins, sugars, tannins, polyphenols, polyalcohols, pectins, lipids, tyrosol, hydroxytyrosol, syringic acid, *p*-hydroxyphenylacetic acid, protocatechuic acid, cinnamic acid, vanillic acid, caffeic acid, *p*-coumaric acid and veratric acid—all of them derived as by-products of olive oil extraction (Chamkha et al., 2001a; Thabet et al., 2004). According to mid-1990s decade reports, approximately 30 million m^3/year of olive mill water contaminate the Mediterranean Basin (Gharsallah, 1994, cited by Chamkha et al., 2001a). Anaerobic digestion is the process through which olive mill wastewaters are treated, and it has been brought to the attention of the scientific community that anaerobic bacteria belonging to the order of Clostridiales are very common in the microbial communities analyzed coming from olive mill water or in the anaerobic digesters used for the treatment of these waters (Thabet et al., 2004).

Among the list of compounds of olive mill wastewaters is cinnamic acid, which is considered of great significance, given its presence in oils and resins of higher plants (Chamkha et al., 2001a). Hydrocinnamic acid is found in the putrefaction products of proteins in soils, and both cinnamic and hydro-cinnamic acids are known to support growth of bacteria in soils. Chamkha et al. (2001a) isolated *Clostridium bifermentans* strain TYR6 from olive mill wastewaters on cinnamic acid. This microorganism is capable of reducing cinnamic acid to hydrocinnamic acid (3-phenylpropionic acid) with or without glucose in the culture. Another microorganism reported regarding its capability of metabolizing cinnamic acid is *Clostridium glycolicum*, also isolated from olive mill wastewater (Chamkha et al., 2001b). Along these lines, other *Clostridium* sp. have been isolated from olive mill wastewaters or the anaerobic digesters treating them, and some of these *Clostridium* sp. have been tested on the biodegradation of aromatics typically found in the olive mill wastewaters; these microorganisms are: *Clostridium xylanovorans* (Mechichi et al., 1999a), *Clostridium methoxybenzovorans* (Mechichi et al., 1999b) and *Clostridium peptidovorans*, capable of fermenting proteins, amino acids and peptides and also of using thiosulfate as terminal electron acceptor (Mechichi et al., 2000) and *Clostridium tunisiense*, which is also a sulfur-reducing bacterium (Thabet et al., 2004). Further work on *Clostridium methoxy-benzovorans* was carried out by Mechichi et al. (2005), who demonstrated the complete mineralization of several poorly oxidizable methoxylated aromatic compounds by coupling the growth of the fermentative *Clostridium* with a nitrate-reducing bacteria: *Thauera* sp. strain Cin3,4.

Ionic Liquids (or "Green"Solvents)

Ionic Liquids (ILs) are a relatively new category of contaminants, given the fact that the scientific community devoted to the biodegradation and bioremediation fields is suggesting that they are a threat to the microbial communities otherwise very appreciated during industrial wastewaters treatment. ILs are being used as "green" solvents because they are non-volatile and recyclable (Rogers et al., 2003). They are currently used in chemical synthesis, both organic and inorganic, in catalysis, also in the recuperation of metals and actinides from aqueous solutions and also in some biological processes (Wang et al., 2011). Compounds of this type are Methoxyethyl-methyl- Imidazolium-based ionic liquids, being this the cationic part of the ionic liquid to which several anions can be attached. Some of the anions that can be attached to it could be tetrafluoroborate $(BF_4)^-$, hexafluorophosphate $(PF_6)^-$, trifluoroacetate $(CF_3COO)^-$, methane sulfonate (OMS), bis (trifluoro-methane) sulfonamide $(Tf_2N)^-$, 1-butyl-3-methyl imidazolium tetrafluoroborate (BMIM)(BF_4).

Due to the recent concern on the damage that these compounds may cause to the environmental microbiology, researchers are conducting studies on the toxicity of the green solvents on certain microorganisms and organisms (Azimova et al., 2006; Ochoa-Herrera et al., 2008; Pham et al., 2008). In general, it has already been demonstrated that the fluorinated compounds are more toxic to the bacteria than the non-halogenated ones (Wang et al., 2011).

Wang et al. (2011) carried out experimental work on toxicity of several ILs on *Clostridium* sp. (ATCC 53464), because it is a microorganism capable of reducing Fe(III) to Fe(II), Mn(IV) to Mn (II), TC(VII) to Tc(IV), U(VI) to U(IV) and Pu(IV) to Pu(III) and also because Clostridia are ubiquitous in the environment and also are found and taken advantage of in many biodegradation and wastewater treatment processes, which makes them a valuable kind of bacteria to take care of. The results of this research group demonstrated that the higher the number of fluorine atoms in the IL, the higher the inhibition of metabolism and growth of the *Clostridium*, a fact that will have to be considered if some engineered process to treat streams contaminated with this type of compounds is required.

Clostridium Sp. and Potential Biotechnological Applications Currently Under Study

Being that this is a chapter basically focused on a review on *Clostridium* sp. and their involvement on the most important biodegradation processes, we certainly did not do a survey on the relatively new enzymes identified for potential biotechnological applications; however it was very noticeable that most of the articles issued for new *Clostridium* are descriptive, and therefore an application of the strains is not included yet. It is also noticeable that nowadays new bacteria belonging to the genus *Clostridium* are still being isolated and classified, whereas some of them such as *Clostridium acetobutylicum* ATCC 824 are known and used for biobutanol production and also studied in experiments of genetic manipulation to obtain "improved" strains of *Clostridium* that can offer higher yields in the production of desired solvents (Green, 2011). In fact, application of *Clostridium* sp. in biofuels (such as biobutanol) production is a subject of study that well would deserve an entire book itself.

In this section, some of the most recent research on *Clostridium* and potential applications will be briefly mentioned in order to inform the reader on the current topics regarding these microorganisms. This includes a little on electron transport mechanisms, thermotolerant enzymes, enzymes isolated from *Clostridium* inhabiting cold environments and finally, relatively new classified *Clostridium* inhabiting warmer environments such as hydrothermal vents and their involvement in some biodegradation processes.

Certainly, *Clostridium acetobutylicum* is one of the most studied strains, thanks to its reputation as useful fermentative bacteria yielding energy and biofuels and the interest of the microbiologists and engineers on the capability of some *Clostridium* to survive under microoxic environments (Honda-Malca et al., 2011).The oxygen detoxification in these *Clostridium* strains has been associated with several proteins such as superoxide dismutase, superoxide reductase, catalase, NAD(P)H-dependent alkyl hydroperoxide reductase and NAD(P)H oxidase, heme-oxygenase, heme-iron oxidoreductases (i.e. cytochromes P450), just to mention a few (Girhard et al., 2007; Hannemann et al., 2007; McLean et al., 2005). In order to clarify the role of flavodoxins and flavodoxin oxidoreductases in oxygen detoxification and electron transfer mechanism, Honda-Malca et al. (2011) conducted experimental work to further explain these proteins' role and gain more information on the electron transfer mechanism in these microorganisms. We would like to refer the reader to the work of these authors, since it may provide useful information to the scientific community dedicated to design and optimization of energy or biofuels production processes.

On the other hand, interesting enzymes expressed by *Clostridium* are still being discovered, and it may be safe to say are to be discovered yet. An example is the novel keratinase (28.7 KDa molecular weight), produced by the thermotolerant *Clostridium sporogenes* bv. pennavorans bv. nov., a name given due to the unique characteristic of this microorganism of biodegrading chicken feathers. This enzyme was active in a range of temperature of 40-70 °C, being the optimal temperature 55°C. The pH range in which the keratinase showed to be active was of 6.5-10. The optimal pH for the enzymatic activity was around 8.0. The purified enzyme showed a specific activity of 623.2 Unit/mg. The unit of keratinase activity was defined as the amount of the enzyme that results in an increase in absorbance at 595 nm of 0.01 after reaction with keratine azure at 50 °C and pH 8.0 for one hour. (Ionata et al., 2008).

Cold environments are also a great source of microorganisms; these include bacteria, archaea, yeasts, filamentous fungi and algae and are distributed in the aquatic cold zones on earth such as the atmosphere and clouds, snow, cryoconite holes, glaciers, Polar and Alpine lakes, deep sea and ice sea. Terrestrial cold environments hosting these microorganisms are cold soils, Alpine soils, Antarctic soils and permafrost (Margesin and Miteva, 2011). Applications of these may be quite a few, quoting Alam et al. (2005), "the cold active anaerobic bacteria producing extracellular proteases can be useful for biodegradation of organic wastes rich in protein under low temperatures." Other applications of these proteases are in food, detergents (cold wash usage) and leather industries. These authors characterized an extracellular protease (46 KDa molecular weight) produced by *Clostridium* sp. SPA3, a microorganism closely related to *Clostridium subterminale* isolated from the lake sediment of Antarctica. The protease was active in a range of temperature of 0-45°C. The optimal protease activity temperature was 37°C, despite the fact that it was isolated from a psycrophilic microorganism (maximum growth referred to cell mass between 5 and 10°C). The ideal pH for the enzyme activity was 7, but it retained 60-70% of its activity at pH 5 and 10. Also, this protease retained more than 80% of its activity when exposed to temperatures as high as 40 and 50°C for 30 min. The evaluation of the inhibition by metals showed that at 10 mM, Hg, Co and Zn exert negative effect on the protease effectiveness.

Certainly, it has been important to highlight the role of the genus *Clostridium* in the main biodegradation process known so far for some of the most important organic pollutants (mostly anthropogenic), but there are cases in which the role of relatively new *Clostridium* sp. needs to be further investigated.

As stated by Guerrero-Barajas et al. (2011), it is important to highlight that among the approximately 150 named species of the genus *Clostridium*, there are only few isolates that have been obtained from marine environments around the world (Oren, 2005). On the one hand, in marine sediments, only less than ten members of the genus *Clostridium* have been isolated: *Clostridium oceanicum* (Smith, 1970), *Clostridium litorale* (Fendrich et al., 1990), *Clostridium halophilum* (Fendrich et al., 1990), *Clostridium grantii* (Mountfort et al., 1994), *Clostridium ganghwense* (Kim et al., 2006) and *Clostridium aestuarii* (Kim et al., 2007). Furthermore, within hydrothermal vents, there have been reported only two *Clostridium* species: *Clostridium caminithermale* (Brisbarre, 2003) and *Clostridium tepidiprofundi* (Slobodkina et al., 2008).

In this regard, it is important to mention that *Clostridium* sp. from the sea, more specifically from hydrothermal vents, and even more specifically from hydrothermal vents sediments, have been poorly studied for biodegradation purposes other than metals precipitation, as previously mentioned in this chapter. At this point, it is worth mentioning that there is work going on currently on the application of consortia from hydrothermal vents sediments enriched to biodegrade highly harmful compounds such as TCE, compound previously mentioned in this chapter.

The work carried out by Guerrero-Barajas et al. (2011) reported that within the microbial community responsible for biodegrading TCE (300 μM) and cultured under sulfate-reducing conditions, there was a clone (PMV01), with 97% of similarity with *Clostridium ganghwense* along with members of the genus *Bacillus* and *Desulfuromonas*. Although there is relatively little information on *Clostridium* in the vents, this genus has been widely studied in its role in fermentation within consortia enriched under sulfate-reducing conditions in engineered systems (Zhao et. al., 2008; Thabet et al., 2009) and biodegradation as is the case of *Clostridium bifermentans* DPH-1, capable of degrading tetrachloroethylene (PCE), TCE, dichloroethenes (*cis* and *trans* 1,2- dichloroethenes), dichloroethanes, vinyl chloride and some other trichloroethanes (Chang et. al., 2000; Okeke et al., 2001), which suggests that there may be other *Clostridium* strains capable of biodegrading volatile organic compounds (VOCs).

Conclusion

In this chapter, the role of *Clostridium* sp. on biodegradation of some priority pollutants and recent trends of energy generation was revised. As it was shown, ubiquitous *Clostridium* strains are involved in biodegradation of azo-dyes, explosives, chlorinated compounds and reduction of metals.

These bacteria hold both the possibility of being taken advantage of in a wide variety of cometabolic processes as well as in direct, enzymatic ones regarding biodegradation and biotransformation of metals. The literature survey suggests that cometabolic processes and syntrophies involving *Clostridium* are of upmost importance in biodegradation of the pollutants listed in this chapter and that there are metabolic characteristics of them that need further study. Certainly, bacteria belonging to this genus are not the only microorganisms that, acting either in consortium or by themselves, may offer an alternative for biodegradation, but it comes to our attention that they enhance and to some degree ensure the success of some of the most important processes occurring in our environment such as acid mine drainage (AMD) and reduction of metals, which are a main environmental concern worldwide.

Some of the remarkable features that can be drawn from the review are the enzymes that can express in the presence of various toxic compounds including dyes, explosives, chlorinated aliphatics and metals. Chemotaxis exhibited in the presence of certain explosives and the response to the different redox mediators are, without a doubt, very insightful mechanisms that may be further applied in engineered bioremediation systems or in any other original bioremediation strategy.

Worth mentioning is that some *Clostridium* sp. are competitive with other bacteria considered electrochemically very active, such as, for example, *Shewanella* sp.; therefore, it is not surprising that currently *Clostridium* is one of the main genus applied in the generation of electricity in MFCs.

Keeping in mind that there are emerging pollutants and still unknown *Clostridium* in several locations in the planet, new capabilities in the *Clostridium* metabolism including new enzymes, polymers and survival mechanisms are still to be discovered.

Taken as a whole, these ubiquitous microorganisms represent a valuable tool in environmental biotechnology in which, so far, they have shown to be remarkable. This survey invites us to maintain the efforts on the study of the *Clostridium* genus and to expect the surprising news that these microorganisms will still bring to the scientific community in the future.

ACKNOWLEDGMENTS

Thanks to Professor Jalal Hawari, Environmental Biotechnology, Biotechnology Research Institute, National Research Council of Canada, Montreal, Canada, for his kind support and to John Wiley & Sons for the license number: 2696760919267. Thanks to Elsevier for the license number: 2696771209845. Thanks to Professor Claudio Garibay Orijel for his suggestions. Also thanks to ICYT-DF for the financial support provided (grant ICYT-PICS-08-79, Mexico).

REFERENCES

Abraham, W.R., Nogales, B., Golyshin, P.N., Pieper, D.H., Timmis, K.N. (2002). Polychlorinated biphenyl-degrading microbial communities in soils and sediments. *Current Opinion in Microbiology*, 5(3), 246-253

Ahmad, F. and Hughes, J.B. (2000). Anaerobic transformation of TNT by *Clostridium*. In: Spain, J.C., Hughes, J.B., Knackmuss, H-J. (eds). *Biodegradation of nitroaromatic compounds and explosives*. Lewis, New York, pp. 185–212

Akob, D.N., Mills, H.J., Gihring, T.M., Kerkhof, L., Stucki, J.W., Anastácio, A.S., Chin, K-J., Kùsel, K., Palumbo, A.V., Watson, D.B., Kotska, J.E. (2008). Functional Diversity and Electron Donor Dependence of Microbial Populations Capable of U(VI) Reduction in Radionuclide-Contaminated Subsurface Sediments. *Applied and Environmental Microbiology*, 74(10):3150-3170.

Alam, S.I., Dube, S., Reddy, G.S.M., Bhattacharya, B.K., Shivaji, S., Singh, L. (2005). Purification and characterization of extracellular protease produced by *Clostridium* sp. from Schirmacher oasis, Antarctica. *Enzyme and Microbiology Technology,* 36:824-831.

Alexandrino, M., Macías, F., Costa, R., Gomes, N.C.M., Canário, A.V.M., Costa, M.C. (2011). A bacterial consortium isolated from an Icelandic fumarole displays exceptionally high levels of sulfate reduction and metals resistance. *Journal of Hazardous Materials*, 187:362-370.

Azimova, M.A., Morton, S.A., Frymier, P.D. (2006). Toxicity of imidazolium-based ionic liquids to industrial wastewater treatment bacteria. Bioengineering Conference, 2006. In*: Proceedings of the IEEE 32nd Annual Northeast*. pp. 11–12.

Bedard, D.L., Bailey, J.J., Reiss, B.L., Jerzak, G.V.S. (2006). Development and Characterization of Stable Sediment-Free Anaerobic Bacterial Enrichment Cultures that Dechlorinate Aroclor 1260. *Applied and Environmental Microbiology*, 72(4): 2460-2470.

Bhushan, B., Paquet, L., Spain, J.C., Hawari, J. (2003). Biotransformation of 2,4,6,8,10,12-hexanitro-2,4,6,8,10,12-hexaazaisowurtzitane (CL-20) by denitrifying *Pseudomonas* sp. strain FA1. *Applied and Environmental Microbiology* 69 (9), 5216e5221.

Bhushan, B., Halasz, A., Thiboutot, S., Ampleman, G., Hawari, J. (2004). Chemotaxis-mediated biodegradation of cyclic nitramine explosives RDX, HMX, and CL-20 by *Clostridium* sp. EDB2. *Biochemical and Biophysical Research Communications*, 316:816-821.

Bhushan, B., Halasz, A., Hawari, J. (2006). Effect of iron (III), humic acids and anthraquinone-2,6-disulfonate on biodegradation of cyclic nitramines by *Clostridium* sp. EDB2. *Journal of Applied Microbiology*, 100:555-563.

Boonchayaanant, B., Nayak, D., Du, X., Criddle, C.S. (2009). Uranium reduction and resistance to reoxidation under iron-reducing and sulfate-reducing conditions. *Water Research*, 43: 4652-4664.

Brisbarre, N., Fardeau, M-L., Cueff, V., Cayol, J-L., Barbier, G., Cilia, V., Ravot, G., Thomas, P., Garcia, J-L., Ollivier, B. (2003*). Clostridium caminithermale* sp. nov., a slightly halophilic and moderately thermophilic bacterium isolated from an Atlantic deep-sea hydrothermal chimney. *International Journal of Systematic and Evolutionary Microbiology,* 53:1043-1049.

Chamkha, M., Patel, B.K.C., Garcia, J.L., Labat, M. (2001a). Isolation of *Clostridium bifermentans* from Oil Mill Wastewaters Converting Cinnamic Acid to 3-phenylpropionic Acid and Emendation of the Species. *Anaerobe*, 7:189-197.

Chamkha M, Labat M, PatelBKC, Garcia JL. (2001b). Isolation of a cinnamic acid-metabolizing *Clostridium glycolicum* strain from oil mill wastewaters and emendation of the species description. *International Journal of Systematic and Evolutionary Microbiology*, 51:2049–2054.

Chang, Y.C., Okeke, B.C., Hastu, M., Takamisawa, K. (2001). In vitro dehalogenation of tetrachloroethylene (PCE) by cell-free extracts of *Clostridium bifermentans* DPH-1. *Bioresource Technology*, 78:141-147.

Chiu, T-C., Yen, J-H., Hsieh, Y-N., Wang, Y-S. (2005). Reductive transformation of dieldrin under anaerobic sediment culture. *Chemosphere*, 60:1182-1189.

Crocker, F.H., Indest, H.L., Tredrizkson, H.L. (2006). Biodegradation of the cyclic nitramine explosives RDX, HMX and CL-20. *Applied Microbiology and Biotechnology*, 73:274-290.

Dario, A., Schroeder, M., Nyanhongo, G.S., Englmayer, G., Guebitz, G.M. (2010). Development of a biodegradable ethylene glycol dinitrate-based explosive. *Journal of Hazardous Materials*, 176:125–130

Desvaux, M. (2005). *Clostridium cellulolyticum*: model organism of mesophilic cellulolytic clostridia. *FEMS Microbiology Reviews*, 29:741-764.

Du, Z., Li, H., Gu, T. (2007). A state of the art review on microbial fuel cells: A promising technology for wastewater treatment and bioenergy. *Biotechnology Advances*, 25:464-482.

Ederer, M. M., Lewis, T. A., Crawford, R. L. (1997). 2,4,6-Trinitrotoluene (TNT) transformation by *Clostridia* isolated from a munition-fed bioreactor: comparison with non-adapted bacteria. *Journal of Industrial Microbiology and Biotechnology*, 18:82–88.

Ekici, P., Leupold, G., Parlar, H. (2001). Degradability of selected azo dye metabolites in activated sludge systems. *Chemosphere*, 44: 721–728.

Esteve-Núñez, A., Caballero, A., Ramos, J.L. (2001). Biological Degradation of 2,4,6-Trinitrotoluene. *Microbiology and Molecular Biology Reviews*, 65(3): 335-352.

Field, J. A. and Brady, J. (2003). Riboflavin as a redox mediator accelerating the reduction of the azo dye mordant yellow 10 by anaerobic granular sludge. *Water Science and Technology*, 48:187-193.

Field, J.A. and Sierra-Alvarez, R. (2004). Biodegradability of chlorinated solvents and related chlorinated aliphatic compounds. *Reviews in Environmental Science & Bio/Technology*, 3:185-254.

Field, J.A. and Sierra-Alvarez, R. (2008a). Microbial degradation of chlorinated dioxins. *Chemosphere*, 71:1005-1018.

Field, J. A. and Sierra-Alvarez, R. (2008b). Microbial transformation and degradation of polychlorinated biphenyls. *Environmental Pollution*, 155: 1-12.

Finch, A.S., Mackie, T.D., Sund, C. J., Sumner, J.J. (2011). Metabolite analysis of *Clostridium acetobutylicum*: Fermentation in a microbial fuel cell. *Bioresource Technology*, 102:312-315.

Forgacs, E., Cserháti, T., Oros, G. (2004). Removal of synthetic dyes from wastewaters: a review. *Environment International*, 30:953-971.

Franciscon, E., Zille, A., Dias Guimaro, F., Ragagnin de Menezes, C., Durrant, L.R., Cavaco-Paulo, A. (2009). Biodegradation of textile azo dyes by a facultative *Staphylococcus arlettae* strain VN-11 using a sequential microaerophilic/aerobic process. *International Biodeterioration and Biodegradation*, 63:280-288.

Franks, A.E. and Nevin, K.P. (2010). Microbial Fuel Cells, A Current Review. *Energies,* 3:899-919.

Fendrich, C., Hippe, J., Gottschalk, G. (1990). *Clostridium halophilium* sp. nov. and *C. litorale* sp. nov., an obligate halophilic and a marine species degrading betaine in the Stickland reaction. *Archives of Microbiology*, 154:127-132.

Furukawa, K. and Fujihara, H. (2008). Microbial Degradation of Polychlorinated Biphenyls: Biochemical and Molecular Features. *Journal of Bioscience and Bioengineering*, 105(5):433-449.

Gao, W. and Francis, A.J. (2008). Reduction of Uranium (VI) to Uranium (IV) by Clostridia. *Applied and Environmental Microbiology*, 74(14): 4580-4584.

García-Balboa, C., Chion Bedoya, I., González, F., Blázquez, M.L., Muñoz, J.A., Ballester, A. (2010). Bio-reduction of Fe(III) ores using three pure strains of Aeromonas hydrophila, Serratia fonticola and *Clostridium celerecrescens* and a natural consortium. *Bioresource Technology*, 101: 7864-7871.

Gharsallah, N. (1994). Influence of dilution and phase separation on the anaerobic digestion of olive mill wastewaters. *Bioprocess Engineering*, 10: 29–34

Girhard, M., Schuster, S., Dietrich, M., Durre, P., Urlacher, V.B. (2007). Cytochrome P450 monooxygenase from *Clostridium acetobutylicum*: a new alpha-fatty acid hydroxylase. *Biochemistry and Biophysics Research Community*, 362: 114–119.

Guerrero-Barajas, C. and Field, J.A. (2005). Enhancement of anaerobic carbon tetrachloride biotransformation in methanogenic sludge with redox active vitamins. *Biodegradation*, 16(3):215-228.

Guerrero-Barajas, C., Garibay-Orijel, C., Rosas-Rocha, L.E. (2011). Sulfate reduction and trichloroethylene biodegradation by a microbial community from hydrothermal vents sediments. *International Biodeterioration and Biodegradation,* 65 (1): 116-123.

Green, E.M. (2011). Fermentative production of butanol – The industrial perspective. *Current Opinion in Biotechnology*, 22(3): 337-343.

Hannemann, F., Bichet, A., Ewen, K.M., Bernhardt, R. (2007). Cytochrome P450 systems - biological variations of electron transport chains. *Biochimica et Biophysic Acta*, 1770: 330–344.

Hata, J., Miyata, N., Kim, E-S., Takamisawa, K., Iwahori, K. (2004). Anaerobic degradation of *cis*-1,2-dichloroethylene and vinyl chloride by *Clostridium* sp. strain DC1 isolated from landfill leachate sediment. *Journal of Bioscience and Bioengineering*, 97(3):196-201.

Hawari, J., Beaudet, S., Halasz, A., Thiboutot, S., Ampleman, G. (2000). Microbial degradation of explosives: biotransformation versus mineralization. *Applied Microbiology and Biotechnology*, 54:605-618.

Hou, L.H. and Dutta, S.K. (2000). Phylogenetic characterization of several para- and meta-PCB dechlorinating Clostridium species: 16s rDNA sequence analyses. *Letters in Applied Microbiology*, 30:238-243.

Honda Malca, S., Girhard, M., Schuster, S., Dürre, P., Urlacher, V.B. (2011). Expression, purification and characterization of two *Clostridium acetobutylicum* flavodoxins: Potential electron transfer partners for CYP152A2. *Biochimica et Biophysica Acta*, 1814: 257-264.

Hu, M., Zhang, Y., Wang, Z.G., Jiang, Z., Li, J. (2011). Influence of humic acid on the trichloroethene degradation by *Dehalococcoides*-containing consortium. *Journal of Hazardous Materials*, 190:1074-1078.

Huang, S., Lindahl, P.A., Wang, C., Bennett, G. N., Rudolph, F.B., Hughes J.B. (2000). 2,4,6-Trinitrotoluene reduction by carbon monoxide dehydrogenase from *Clostridium thermoaceticum. Applied and Environmental Microbiology,* 66:1474–1478.

Hughes, J. B., Wang, C., Yesland K., Bhadra, R., Richardson, A., Bennet, G., Rudolph, F. (1998a). Reduction of 2,4,6-trinitrotoluene by *Clostridium acetobutylicum* through hydroxylamino intermediates. *Environ. Toxicol. Chem.* 17:343–348.

Hughes, J. B., Wang, C., Yesland K., Richardson, A., Bhadra, R., Bennet, G., Rudolph, F. (1998b). Bamberger rearrangement during TNT metabolism by *Clostridium acetobutylicum*. *Environmental Science and Technology,* 32:494–500.

Hughes, J. B., Wang, C. Y., Zhang, C. (1999). Anaerobic biotransformation of 2,4-dinitrotoluene and 2,6-dinitrotoluene by *Clostridium acetobutylicum*: a pathway through dihydroxylamino intermediates. *Environmental Science and Technology,* 33:1065–1070.

Jin, X., G. Liu, Z. Xu, Tao, W. (2007). Decolourisation of a dye industry effluent by *Aspergillus fumigatus* XC6. *Applied Microbiology and Biotechnology*, 74: 239

Joe, M.H., Lim, S.Y., Kim, D.H., Lee, I.S. (2008). Decolorization of reactive dyes by *Clostridium bifermentans* SL186 isolated from contaminated soil. *World Journal of Microbiology and Biotechnology*, 24:2221-2226.

Khan, T. A., Bhadra, R., Hughes, J. (1997). Anaerobic transformation of 2,4,6-TNT and related nitroaromatic compounds by *Clostridium acetobutylicum*. *Journal of Industrial Microbiology and Biotechnology*, 18:198–203.

Khelifi, E., Bouallagui, H., Fardeau, M.L., Touhami, Y., Godon, J-J., Cayol, J-l., Ollivier, B., Hamdi, M. (2009). Fermentative and sulphate-reducing bacteria associated with treatment of an industrial dye effluent in an up-flow anaerobic fixed bed bioreactor. *Biochemical Engineering Journal*, 45:136-144.

Kim, D.J, An, H.Y., Yoon, J.H., Park, Y.H., Kawai, F., Jung, C.M. (2002). Identification of *Clostridium perfringens* AB&J and its uptake of bromophenol blue. *Journal of Microbiology and Biotechnology*,12:544–52

Kim, S., Jeong, H., Kim, S., Chun, J. (2006). *Clostridium ganghwense* sp. nov., isolated from tidal flat sediment. *International Journal of Systematic and Evolutionary Microbiology*, 56:691-693.

Kim, S., Jeong, H., Chun, J. (2007). *Clostridium aestuarii* sp. nov., from tidal flat sediment. *International Journal of Systematic and Evolutionary Microbiology,* 57:1315-1317.

Kulkarni, M. and Chaudhari, A. (2007). Microbial remediation of nitro-aromatic compounds: An overview. *Journal of Environmental Management*, 85:496-512

Kutty, R., and Bennett, G. (2005). Biochemical characterization of trinitrotoluene transforming oxygen-insensitive nitroreductases from *Clostridium acetobutylicum* ATCC 824. *Arch Microbiology*, 184:158–167

Lewis, T.A., Newcomb, D.A., Crawford, R.L. (2004). Bioremediation of soils contaminated with explosives. *Journal of Environmental Management*, 70, 291-307.

Lewis, T. A., Goszczynski, S., Crawford, R. L., Korus, R.A., Admassu, W. (1996). Products of anaerobic 2,4,6-trinitrotoluene (TNT) transformation by *Clostridium bifermentans*. *Applied and Environmental Microbiology,* 62:4669–4674.

Lloyd, J.R. (2003). Microbial reduction of metals and radionuclides. *FEMS Microbiology Reviews,* 27:411-427.

McCormick, N.G., Cornell, J.H., Kaplan, A.M. (1981). Biodegradation of hexahydro-1,3,5-trinitro-1,3,5 triazine. *Applied and Environmental Microbiology*, 42:817-823.

McLean, K.J., Sabri, M., Marshall, K.R., Lawson, R.J., Lewis, D.G., Clift, D., Balding, P.R., Dunford, A.J., Warman, A.J., McVey, J.P., Quinn, A.M., Sutcliffe, M.J., Scrutton, N.S., Munro, A.W. (2005). Biodiversity of cytochrome P450 redox systems, *Biochem. Soc. Trans.* 33:796–801.

Madden, A.S., Smith, A.C., Balkwill, D.L., Fagan, L.A., Phelps, T.J. (2007). Microbial uranium immobilization independent of nitrate reduction. *Environmental Microbiology*, 9(9): 2321-2330.

Margesin, R. and Miteva, V. (2011). Diversity and ecology of psychrophilic microorganisms. *Research in Microbiology,* 162:346-361.

Martins, M., Faleiro, M.L., Chaves, S., Tenreiro, R., Santos, E., Costa, M.C. (2010a). Anaerobic bio-removal of uranium (VI) and chromium (VI): Comparison of microbial community structure. *Journal of Hazardous Materials*, 176: 1065-1072.

Martins, M., Faleiro, M.L., Chaves, S., Tenreiro, R., Santos, E., Costa, M.C. (2010b). Effect of uranium (VI) on two sulphate-reducing bacteria cultures from a uranium mine site. *Science of the Total Environment*, 408, 2621-2628.

Mechichi, T., Labat, M., Garcia, J.L, Thomas, P., Patel, B.K.C. (1999a). Characterization of a new xylanolytic bacterium *Clostridium xylanovorans* sp. nov. *System Appl Microbiology*, 22:366–371.

Mechichi, T., Labat, M., Patel, B.K.C., Woo, T.H.S., Thomas, P., Garcia, J.L. (1999b). *Clostridium methoxybenzovorans* sp. nov. a new aromatic o-demethylating homoacetogen from an olive mill wastewater treatment digester. *International Journal Systematic and Evolutionary Microbiology,* 49:1201 –1209.

Mechichi, T., Fardeau, M.L., Labat, M., Garcia, J.L., Verhe′ F., Patel, B.K.C. (2000). *Clostridium peptidovorans* sp. nov., a peptide-fermenting bacterium from an olive wastewater treatment digester. *International Journal of Systematic and Evolutionary Microbiology,* 50:1259–1264.

Mechichi, T., Patel, B.K.C., Sayadi, S. (2005). Anaerobic degradation of methoxylated aromatic compounds by *Clostridium* methoxybenzovorans and a nitrate-reducing bacterium *Thauera* sp. strain Cin3,4. *International Biodeterioration and Biodegradation,* 56:224-230.

Mountfort, D.O., Rainey, F.A., Burghardt, J., Stackebrandt, E. (1994). *Clostridium grantii* sp. nov., a new obligately anaerobic, alginolytic bacterium isolated from mullet gut. *Archives of Microbiology*, 162:173-179.

Morris, J.M. and Jin, S. (2008). Feasibility of using microbial fuel cell technology in bioremediation of hydrocarbons in groundwater. *Journal of Environmental Science and Health, Part A: Environmental Science Engineering*, 43: 18–23.

Morris, J.M., Jin, S., Crimi, B., Pruden, A. (2009). Microbial fuel cell in enhancing anaerobic biodegradation of diesel. *Chemical Engineering Journal*, 146:161-167.

Nyanhongo, G.S., Aichernig, N., Ortner, M., Steiner, W., Guebitz, G.M. (2009). Incorporation of 2,4,6-trinitrotoluene (TNT) transforming bacteria into explosive formulations. *Journal of Hazardous Materials*, 165:285–290.

Niessen, J., Schröder, U., Scholz, F.(2004). Exploiting complex carbohydrates for microbial electricity generation – a bacterial fuel cell operating on starch. *Electrochemistry Commun.* 6, 955–958.

Ochoa-Herrera, V., Banihani, Q., Leon, G., Khatri, C., Field, J.A., Sierra-Alvarez, R. (2009). Toxicity of fluoride to microorganisms in biological wastewater treatment systems. *Water Research*, 43: 3177–3186.

Ohisa, N., Yamaguchi, M., Kurihara, N., 1980. Lindane degradation by cell-free extracts of *Clostridium rectum*. *Arch. Microbiology*, 25 (3): 221–225.

Okeke, B., Chang, Y.C., Hatsu, M., Suzuki, T., Takamizawa, K. (2001). Purification, cloning, and sequencing of an enzyme mediating the reductive dechlorination of tetrachloroethylene (PCE) from *Clostridium bifermentans* DPH-1. *Canadian Journal of Microbiology*, 47(5):448-456.

Osman, M.H., Shah, A.A., Walsh, F.C. (2010). Recent progress and continuing challenges in bio-fuel cells. Part II: Microbial. *Biosensors and Bioelectronics,* 26:953-963.

Pandey, A., Singh, P., Iyengar, L. (2007). Bacterial decolorization and degradation of azo dyes. *International Biodeterioration and Biodegradation,* 59:73-84.

Pant, D., Van Bogaert, G., Diels, L., Vanbroekhoven, K. (2010). A review of the substrates used in microbial fuel cells (MFCs) for sustainable energy production. *Bioresource Technology*, 101:1533-1543.

Pham, T.P.T., Cho, C.W., Min, J., Yun, Y.S. (2008). Alkyl-chain length effects of imidazolium and pyridinium ionic liquids on photosynthetic response of *Pseudokirchneriella subcapitata*. *Journal of Bioscience and Bioengineering,* 105. 425–428.

Pecheritsyna, S.A., Arkhipova, O.V., Suzina, N.E., Lysanskaya, V.Y., Laurinavichyus, K.S., Shcherbakova, V.A. (2011). Intracellular polysaccharide of an anaerobic psychrophile *Clostridium algoriphilum*. *Microbiology*, 80(1): 37-42.

Quintero, J.C., Moreira, M.T., Feijoo, G., Lema, J.M. (2005). Anaerobic degradation of hexachlorocyclohexane isomers in liquid and soil slurry systems. *Chemosphere*, 61:528-536.

Rabaey, K., Boon, N., Siciliano, S.D., Verhaege, M., Verstraete, W. (2004). Biofuel Cells Select for Microbial Consortia That Self-Mediate Electron Transfer. *Applied and Environmental Microbiology*, 70(9): 5373-5382.

Rabaey, K. and Verstraete, W. (2005). Microbial fuel cells: novel biotechnology for energy generation. *Trends in Biotechnology*, 23(6): 291-298.

Ren, Z., Steinburg, L.M., Regan, J.M. (2008). Electricity production and microbial biofilm characterization in cellulose-fed microbial fuel cells. *Water Science Technology*, 58 (3): 617–622.

Rogers, R.D., Seddon, K.R. (2003). *Ionic Liquids as Green Solvents: Progress and Prospects*. Oxford University Press.

Rosser, S.J., Basran, A., Travis, E.R., French C.E., Bruce, N.C. (2001). Microbial transformations of explosives. *Advances in Applied Microbiology*, 49:1-35.

Sagi-Ben-Moshe, S., Ronen, Z., Dahan, O., Weisbrod, N., Groisman, L., Adar, E., Nativ, R. (2009). Sequential biodegradation of TNT, RDX and HMX in a mixture. *Environmental Pollution*, 157: 2231-2238.

Saratale, R.G., Saratale, G.D., Chang, J.S., Govindwar, S.P. (2011). Bacterial decolorization and degradation of azo dyes: A review. *Journal of the Taiwan Institute of Chemical Engineers*, 42(1):138-157.

Sembries, S. and Crawford, R. (1997). Production of *Clostridium bifermentans* spores as inoculum for bioremediation of nitroaromatic contaminants. *Applied and Environmental Microbiology*, 63:2100–2104

Shelley, M.D., Autenrieth, R.L, Wild, J.R., Dale, B.E. (1996). Thermodynamic analysis of trinitrotoluene biodegradation and mineralization pathways. *Biotechnology and Bioengineering*, 51:198–205.

Shen, C. F., Hawari, J.A., Ampleman, G., Thiboutot, S., Guiot, S.R. (2000). Origin of *p*-cresol in the anaerobic degradation of trinitrotoluene. *Canadian Journal of Microbiology*, 46:119–124.

Slobodkina, G.B., Kolganova, T.V., Tourova, T.P., Kostrikina, N.A., Jeanthon, C., Bonch-Osmolovskaya, E.A., Slobodkin, A.I. (2008). *Clostridium tepidiprofundi* sp. nov., a moderately thermophilic bacterium from a deep-sea hydrothermal vent. *International Journal of Systematic and Evolutionary Microbiology,* 58:852-855.

Smith, L.D. (1970). *Clostridium oceanicum*, sp. n., a Spore forming anaerobe isolated from marine sediments. *Journal of Bacteriology*, 103(3): 811-813.

Stenuit, B.A. and Agathos, S.N. (2010). Microbial 2,4,6-trinitrotoluene degradation: could we learn from (bio)chemistry for bioremediation and vice versa?. *Applied Microbiology and Biotechnology*, 88:1043-1064.

Stewart, D.I., Burke, I.T., Hughes-Berry, D.V., Whittleston, R.A. (2010). Microbially mediated chromate reduction in soil contaminated by highly alkaline leachate from chromium containing waste. *Ecological Engineering,* 36:211-221.

Strigul, N., Braida, W., Christodoulatos, C., Balas, W., Nicolich, S. (2006). The assessment of the energetic compound 2,4,6,8,10,12-hexanitro-2,4,6,8,10,12-Hexaazaisowurtzitane (CL-20) degradability in soil. *Environmental Pollution*, 139:353-361.

Sund, C.J. and McMasters, S. (2007). Effect of electron mediators on current generation and fermentation in a microbial fuel cell. *Biotechnological Products and Process Engineering*, 76:561-568.

Suresh, K., Prakash, D., Rastogi, N., Jain, R.K. (2007). *Clostridium nitrophenolicum* sp. nov., a novel anaerobic p-nitrophenol-degrading bacterium, isolated from a subsurface soil sample. *International Journal of Systematic and Evolutionary Microbiology*, 57:1886-1890.

Takahashi, Y., Suto, K., Inoue, C. (2010). Polysulfide reduction by Clostridium relatives isolated from sulfate-reducing enrichment cultures. *Journal of Bioscience and Bioengineering*, 109(4): 372-380.

Thabet, O.B.D., Fardeau, M.L., Joulian, C., Thomas, P., Hamdi, M., Garcia, J.L., Ollivier, B. (2004). *Clostridium tunisiense* sp. nov., a new proteolytic, sulfur-reducing bacterium isolated from an olive mill wastewater contaminated by phosphogypse. *Anaerobe*, 10:185-190.

Trott, S., Nishino, S.F., Hawari, J., Spain, J.C. (2003). Biodegradation of the nitramine explosive CL-20. *Applied and Environmental Microbiology* 69:1871-1874.

Van der Zee, F.P., Burwman, R.H.M., Strik, D.P.T.B., Lettinga, G., Field, J.A. (2001). Application of redox mediators to accelerate the transformation of reactive azo dyes in anaerobic bioreactor. *Biotechnology and Bioengineering.* 75 (6): 691–701.

Vecchia, E.D., Veeramani, H., Suvorova. E.I., Wigginton, N.S., Bargar, J.R., Bernier-Latmani, R. (2010). U(VI) reduction by spores of *Clostridium acetobutylicum. Research in Microbiology*, 161:765-771.

Wang, H., Malhotra, S.V., Francis, A.J. (2011). Toxicity of various anions associated with methoxyethyl methyl imidazolium-based ionic liquids on *Clostridium* sp. *Chemosphere*, 82:1597-1603.

Zanaroli, G., Balloi, A., Negroni, A., Daffonchio, D., Young, L.Y., Fava, F.(2010). Characterization of the microbial community from the marine sediment of the Venice lagoon capable of reductive dechlorination of coplanar polychlorinated biphenyls (PCBs). *Journal of Hazardous Materials,* 178:417-426

Zhao, J.S., Spain, J., Hawari, J. (2003). Phylogenetic and metabolic diversity of hexahydro-1,3,5-trinitro-1,3,5-triazine (RDX)-transforming bacteria in strictly anaerobic mixed cultures enriched on RDX as nitrogen source. *FEMS Microbiology Ecology*, 46:189-196.

Zhao, J.S., Paquet, L., Halasz, A., Manno, D., Hawari, J. (2004). Metabolism of octahydro-1,3,5,7-tetranitro-1,3,5,7-tetrazocine by *Clostridium bifermentans* strain HAW-1 and several other H2-producing fermentative anaerobic bacteria. *FEMS Microbiology Letters*, 237:55-72.

Zhang, C. and Hughes, J.B. (2003). Biodegradation pathways of hexahydro-1,3,5-trinitro-1,3,5-triazine (RDX) by *Clostridium acetobutylicum* cell-free extract. *Chemosphere*, 50:665-671.

In: Clostridia
Editors: S. V. Clinton and M. F. Kelly
ISBN: 978-1-62100-761-6
© 2012 Nova Science Publishers, Inc.

Chapter 2

CLOSTRIDIA: HARVESTING THE POTENTIAL BENEFITS FOR HUMANS

Caroline Hinton[1], Shufeng Zhou[2] and Ming Q. Wei[1]*

[1]Division of Molecular and Gene Therapies, Griffith Health Institute and School of Medical Science, Gold Coast campus, Griffith University, Qld., Australia

[2]Department of Pharmaceutical Sciences, College of Pharmacy, University of South Florida, Bruce B Downs Blvd., Tampa, Florida, US

ABSTRACT

The genus *Clostridium* consists of over 295species of gram-positive, rod shaped bacteria. These are mostly strict anaerobes and present ubiquitously in the environment due to their ability to form heat, radiation and chemical resistant spores. A number of clostridial species also inhabit the gastrointestinal tracts of animals and humans. Most of clostridial species are harmless microbes; some even are well-known for being

---

* Corresponding author: Professor Ming Q Wei, Division of Molecular and Gene Therapies, Griffith Health Institute and School of Medical Science, Griffith University, Gold Coast campus, Southport, Qld., 4222, Australia, Fax: +61 7 55528908 E-mail: m.wei@griffith.edu.au.

employed to benefit humans,but only a few cause diseases that gave the genus a bed name.

For centuries, clostridia are being explored for various industrial purposes, including food processing, production of detergents due to their enzymes' ability in degrading starch, cellulose and collagen as well as many others. Others are being utilised in waste management at a mass industry scale as well as significant use in the development of acetone, butanol, ethanol and hydrogen as environmentally friendly chemicals and fuels.The neurotoxin Botulimum from *Clostridiumbotulinium*, is well known for its use in cosmetic enhancement and pharmaceutical treatment of various neuronal disorders, hyperhidrosis, gastrointestinal disorders, and dermatological disorders. Other Clostridia-derivedtoxins are also being incorporated into drug delivery systems to the nervous system, and used as prevention in food poisoning and inflammatory responses. Clostridia's preference for anaerobic environments and production of spores has potential for use as therapy for solid tumours with complex, but unique hypoxic and necrotic microenvironment. Early studies showed that spore administration of non-pathogenic strains of clostridia can be done safely and lead to the colonisation within the hypoxic/necrotic regions of tumours, resulting in colonisation of the spores and rapid lysis of the solid tumours. With the advancement of genetic engineering, the limitations of using Clostridia for development of chemicals and biofuels, such as low yields due to solvent toxicity and low productivity, are gradually being overcome. Similarly, genetically engineering varying strains of *Clostridium* to enable production of proteins harmful to tumour cells, will improve the efficacy of oncolysis. While these genetically modified clostridia have so far been shown to be highly effective in animals, its effectiveness in humans and effects of intact immune systems of humans is still being elucidated.

In this chapter we will overview the *Clostridium* genus and how the characteristics which define the *Clostridium* have been utilised in numerous industrial, biotechnological and medicalendeavours. We will also discuss the development of clostridial spores as oncolytic agents or "Trojan horse" vectors for cancer gene therapy. Finally, we will look at advances made in the field and evaluate its futurepotentials in cancer gene therapy.

INTRODUCTION

Several species of the *Clostridium* genus have gained fame in the past for their role in harmful diseases such as botulism, tetanus, gas gangrene and pseudomembranous colitis. However, the vast majority of *Clostridium* species are non-pathogenic and have played, and continue to play, a significant role in various industrial processes and medical contexts. With the advancement of genetic engineering and the enhanced knowledge of various clostridial genomes and detailed molecular understanding of the organisms, we will be able to further harvest the many benefits offered by *Clostridium*, whether it is the creation of new-age biofuels, daily use of healthy food or the treatment of fatal cancers.

CLOSTRIDIA: BACKGROUND

The *Clostridium* genus, belonging to the family *Bacillaceae* of the phylum *Firmicutes*, consists of over 295 bacterial species, many of which were newly discovered [1].All *Clostridium* species have particular common properties (Table 1) including anaerobic and fermentative metabolism, gram-positive cell wall structure and the capability to form heat, radiation and chemical resistant endospores [2-4] in an irreversible process, eventually leading to the formation of a viable and vegetative cell[5]. Majority of Clostridium species are rod-shaped ranging from 0.4 to 1.3μM in width and 3 to 8μM in length. Most also have peritrichous flagella, enabling motility [2].

Anaerobic and fermentative metabolism by Clostridia involves the secretion of a number of hydrolytic enzymes capable of degrading a range of organic substrates to provide nutrients and energy required for their survival [6]. Along with spore germination, their diverse metabolic properties, allows their growth in both favourable and unfavourable conditions [5]. They can be found ubiquitously in soil and water sediments, as well as intestinal tract of humans and animals [6, 7]. While most Clostridium species live as free bacteria, a number also exist as pathogens causing disease in humans and animals [8].

Molecular Microbiology of the *Clostridium*

Although a considerable number of clostridial species are of significant medical and biotechnological importance, only six (*C. acetobutylicum, C. difficile, C. perfringens, C. botulinum, C novyi, and C. tetani*) have been genomically sequenced [9-14].Increasingly recognised as important regulatory molecules in all organisms, a number of small RNAs (sRNA) have been identified in the Genus *Clostridium* [15].

Table 1. General Characteristics of *Clostridium spp*

Characteristics of Clostridium		Reference
Oxygen sensitivity	Obligate anaerobe	[6]
Gram stain	positive	[7]
morphology	Rod (0.4 to 1.3μM in width and 3 to 8μM in length)	[2,7]
Spore-forming	yes	[2, 5,7]
G+C content	Low	[9-14]
Appendages	Peritrichous flagella	[2]
Metabolism	Fermentative	[6]
Enzyme production	Variety of extracellular hydrolytic enzymes	[6]
Growth on culture media	Swarm on agar plates	
Pathogenicity	Mostly saprophytic with only a few pathogenic	[8]
Habitat	Soil, water sediment and GI tract of humans and animals	[6,7]

Genomically Sequenced Clostridium Spp

The 3,031,430-bp sequence of *C. Perfringens* strain 13, consists of 2,660 protein encoding regions, 10rRNA genes and a low G+C content (28%). While found to consist of typical anaerobic fermentation enzymes that lead to gas formation and saccharolytic enzymes, no enzymes for tri-carboxylic acid cycle, respiratory chain or amino acid biosynthesis were found. The genome also consisted of 20 new genes identified as putative virulence factors for *C.Perfringens* as well as 5 hyaluronidase genes which also contribute to virulence.Genome sequencing of *C.perfringens* showed that the bacterium obtains essential materials from its host through the production of degradative enzymes and toxins resulting in destruction of host cells [9].

*C.acetobutylicum*ATCC 824 was also genomically sequenced, and found to consist of a 3,940,880bp chromosome with 11 ribosomal operons and a 192kb mega-plasmid that contains the majority of genes used for solvent production. 3740 polypeptide encoding ORFs and 107RNA genes were also identified [10].

C. tetani consists of a 2,799,205 bp chromosome that encodes 2,372 ORFs. The genome also consists of a 74,082-bp plasmid containing 61 ORFs and encoding the tetanus toxin and collagenases. Other virulence factors were also identified including a number of adhesion and surface layer proteins, some being unique to *C.Tetani* [11].

C.novyi-NT genome consists of 2,547,720 base pairs, low G+C content (28.9%), a new type of transposition and 139 genes that do not have homologs in other bacteria [12].

The *C.difficile* genome consists of a 4,290,252 bp chromosome and a 7,881 bp plasmid. Both the chromosome and plasmid contain a low G+C content of 29.06% and 27.9%, respectively. While the chromosome was found to encode 3,776 predicted coding sequences, the plasmid contains 11 coding sequences, none of which has any obvious function. Only 567 coding sequences were found to be shared with the 4 other sequences *Clostridium* (*C.acetobutylicum, C.botulinum, C.perfringens* and *C.tetani*), while 1,893 are unique to *C*.difficile [13].

The last *Clostridium* genome to be sequenced is that of*C. botulinum* strain Hall A. This genome was found to consist of a 3,886,916bp chromosome and a 16,344bp plasmid containing 3650 and 19 genes respectively. A large number of these genes encode secreted proteases and enzymes involved in metabolism as well as the ability to degrade chitin. The genome indicates *C.botulinum* relies on its toxin to kill various prey species in order to gain nutrients. The genome also shows it is adapted to saprophytic lifestyle in soil and water[14].

Small RNAs in Clostridium

Small RNAs play important regulatory roles in various cellular processes, typically acting as post - transcriptional regulators by interacting with mRNA, modulating target stability and/or translation initiation [16]. Approximately 50 to 500 nucleotides in length [17], sRNAs in *Clostridium* remained unexplored until recently when sRNAs were predicted in 21 clostridial genomes using comparative genomics analyses combined with predications of rho-independent terminators and promoters [15]. This research was able to produce a list of identified and verified sRNAs, with the intention of accelerating interest and further research on sRNAs in *Clostridium*.

Toxins and Disease

Pathogenic Clostridia are not invasive,only producing symptoms of disease through the release of toxins. Of the 295 species of Clostridium, only approximately 15 are known to be capable of producing toxins responsible for disease in humans and animals [18].Clostridial toxins consist of a wide range of biologically active proteins that are antigenic in nature [19] and differing in structure, size, mode and site of action, and ability to diffuse in the organism

[6]. Whilst some toxins act locally, others, for example botulinum, are capable of passing through the mucosal barrier and be distributed to other tissues via the circulatory system [6]. Clostridial toxins also vary in the type of cells they interact with. While the vast majority are involved in gangrene, other act on more specific cells such as intestinal cells, causing necrotic enteritis and other intestinal disease, and neuronal cells causing neurological disease such as tetanus and botulism [6].

Major toxins produced by various *Clostridium spp.* include α-toxin, botulism neurotoxin A-F, β-hemolysin, enterotoxin,ε toxin, τ- toxin and collagenases (Table 2). α-toxin(Cpa) is the only toxin produced by *C. septicum.* The release of α-toxin by *C.septicum* results in gas gangrene and necrotising enterocolitis in humans and Braxy in cattle and sheep [20]. β-hemolysin produced by *C.haemolyticum* is also a phospholipase C, lysing hepatocytes and erythrocytes leading to hemoglobinuria [21]. One well known toxin is the Botulinum neurotoxin (BoNT). Produced by *C.botulinum*, this toxin acts by inhibiting acetocholine release after binding on motor nerve endings causing flaccid paralysis[22].Another well-known neurotoxin is tetanus neurotoxin (TeNT) produced by *C.tetani*. This toxin blocks neurotransmitter release at inhibitory interneurons of the spinal cord causing spastic paralysis [23].

C.perfringens is capable of releasing a number of toxins. One such toxin is α-toxin causing gas gangrene and puerperal septicaemia in humans. CPA is a phospholipase C that hydrolyses the phosphodiester bonds in the main phospholipids of mammalian cells; phosphatidykcholine and sphingomyelin[24]. Another toxin produced by *C.perfringens* is the *C. perfringens* Enterotoxin (CPE), a single polypeptide, which is associated with food poisoning and antibiotic-associated diarrheoa [25]. Others include β- toxin which forms pores in cell membranes leading to swelling and cell lysis [26], ε toxin responsible for enterotoxemia in a number of livestock animals [27] by increasing permeability of the small intestine[28] after binding to a specific, yet unknown receptor [29] andlastly, the τ- toxin, a member of the binary toxin family along with *C.spiroforme* τ- toxin, *C.difficile* toxin (CDT) and *C.botulinum* C2 toxin, which has cytotoxic and enterotoxic properties as well as inducing intestinal histo-pathological damage [30].

Pathogenic *Clostridium*infects their host by one of two means. The first is the oral route, where the bacteria are ingested colonising parts of the digestive tract and releasing toxins resulting in the onset of intestinal disease (Table 2). The second route is via existing wounds where the bacteria grow in the presence of necrotic tissues and produce toxins resulting in gas gangrene or other neurological disease (Table 2).

Table 2. Clostridial toxins; entry and disease

Clostridium Species	Toxin	Entry	Resulting Disease	Reference
Toxic *C.baratii*	Rarely BoNT C and F	Oral	Botulism: human	[31]
C.botulinum	Botulinum neurotoxin A and B (BoNTA/B) BoNT/A,B, E	Wound Oral	Botulism: humans Botulism: humans	[8,22,32]
Toxic *C.butyricum*	BoNT/C, D	Oral	Botulism: anima (avian, bovine, ovine, equine)	[33, 34, 35]
C.chauvoei		Wound	Blackleg: bovine, sheep	[36, 19]
C.difficile	C.difficile toxin A and B (TcdA/B), C.difficile transferase (CDT)	Oral	Pseudomembranous colitis: humans Postantibiotic diarrhea: humans Diarrhoea: pigs, calves Enterocolitis: horses	[8, 37, 38]
C.haemolyticum	β-hemolysin	Wound	Bacillary hemoglobinuria: bovine, sheep	[19, 36]
C.histolyticum	Collagenases	Wound	Gangrene: animals, humans	[19, 36]
C.novyi	α-toxin	Wound	Gangrene: humans, animals Necrotic hepatitis: animals	[19, 36]
C. perfringens	α-toxin Enterotoxin	Wound Oral	Gangrene: Humans, animals Puerperal septicaemia: humans	[8, 24]

Table 2. (Continued)

Clostridium Species	Toxin	Entry	Resulting Disease	Reference
	β	Oral	Food intoxication: humans Sporadic diarrhea: humans, pigs post-antibiotic diarrhea: humans SIDS: humans (some cases)	[25] [26]
	β2	Oral	Necrotic enteritis: humans, young animals Enterotoxemia: Sheep	 [39]
	ε	Oral	Necrotic enteritis: piglets Colitis: horses	
	τ	Oral	Enterotoxemia: sheep, goat	[27]
			Enterotoxemia: calves, rabbit Enteritis: sheep	 [8,30]
	Necrotic enteritis toxin B-like (NetB)	Oral	Necrotic enteritis: birds	[40]
C.septicum	α-toxin		Gangrene: humans, animals Intestinal myonecrosis: humans Braxy: cattle, sheep	[20]
C. sordellii	*C.sordellii* lethal toxin (TcsL), *C.sordellii* hemorrhagic toxin (TcsH)	Oral	Gangrene: Humans, animals Gangrene: humans, animals Enterotoxaemia: sheep, cattle Hemorrhagic enteritis: sheep, cattle	[41, 42]
C. spiroforme	*C.spiroforme* toxin	Oral	Diarrhoea: rabbit	[43]
C. tetani	Tetanus neurotoxin (TeNT)	Wound	Tetanus: Humans, animals	[8, 23, 44, 45]

Industrial Beneficial Clostridium

Dating back to ancient times, *Clostridium* species have been utilised for many biotechnological processes. Indigo formation by *C. isatidis* [2], was used for dyeing cloth in the times of the Iron and Middle Ages and in between [46]. More recently, clostridial enzymes have been an important part of food processing, food production in detergents [2] and the anaerobic digestion of organic waste materials [47]. Fermentation properties of *Clostridium* have been vital in the production of bio-fuels and bio-chemicals. With the progress of genetic engineering, improved production of such vital biofuels and chemicals has been, and continues to be, achieved.

Food Production, Food Processing and Organic Waste Digestion

Enzymes used in food production and food processing include cellulases, amylases, pectinases, proteinases, collagenases and numerous others. Species that produce these enzymes include *C.thermocellum, C.thermosulfurogenes, C.thermohydrosulfuricum and C.thermosaccharolyticum* [2].

Clostridium species are also involved in the anaerobic digestion of waste products. While they are the most common among the degraders in anaerobic conditions [48], anaerobic digestion of waste products rarely relies on a single microbial strain but on a number of different microbes [49]. Using a series of metabolic reactions such as hydrolysis, acidogenesis and methogensis [50], anaerobic digestion is the process of biologically transforming an organic substance, such as landfill waste, into another form, such as biogas and other energy-rich organic compounds [51] in the absence of oxygen. Chemical compounds created by the anaerobic digestion of organic waste have the potential for use as fuel, easing dependence on fossil fuel for energy [52]. This process is an example of renewable biomass.

Biofuel and Chemical Production by Clostridia- New Age Energy?

A variety of microorganisms, including various *Clostridium* species, are currently being utilised in the production of chemicals and biofuels from renewable biomass. These include higher alcohols [53-55], biodiesel [56], alkanes [57] and hydrogen [58].

Ethanol and butanol, however, are the most commonly used bio-energy compounds [59]. In Strict anaerobic conditions, *Clostridium spp.* is capable of producing butanol, along with smaller amounts of acetone and ethanol, from renewable biomass in a process called the acetone-butanol-ethanol (ABE) fermentation process [60]. This process was first reported by Louis Pasteur in 1861. It wasn't until early 20th century that ABE fermentation by microorganisms was done at an industrial production level [61]. This came after Chaim Weizmann discovered *C. acetobutylicum*, which naturally produces acetone, butanol and ethanol (ABE) in a ratio of 3:6:1 [62]. ABE fermentation plants were opened during the two world wars in US, UK, China, Russia, South Africa and India due to the demand for acetone, while butanol was an unwanted by-product. After the war, butanol became a higher importance [60,63]. However, ABE yield is often low due to the low efficiency of Clostridium to convert substrate into ABE [59] making ABE fermentation costly. During the 1950's, the focus on bio-energy declined when petrochemical-based production of butanol and ethanol provided a cheaper alternative [60, 63, 64]. Therefore, research continues to perfect the fermentative process and reduce costs associated with clostridial ABE fermentation.

With the depleting availability of fossil fuels and increasing concerns of greenhouse gasses causing climate change [65], there has been a recent re-focus on bio-chemical production by *Clostridium*. ABE fermentation from renewable biomass by *Clostridium spp*. has been one of the most attractive sources of bio-chemicals. Not only are all three solvents used in many industrial applications, but, in theory, it has the potential to completely replace petro-chemically derived energy [60]. In comparison to ethanol, butanol as a biofuel is a more favourable option, due to its lower vapour pressure, higher energy content, higher boiling point and a gasoline like octane rating, permitting it to be blended with current gasoline at higher proportions than ethanol [59, 66], eventually being able to replace gasoline altogether [63]. Butanol can be used in numerous other applications in other fields including plastics and food production [59]. Therefore, optimizing butanol production is an attractive approach commercially. When a strain of Clostridia is available that is not only able to selectively produce a high yield of butanol but able to ferment polymeric carbohydrates of a low value such as biomass, butanol production by bacteria may become cost effective again compared to that of petro-chemical production [67].

Saccharolytic mesophilic species of clostridia capable of forming butyrate are the only species able to produce butanol [68]. These species include *C. acetobutylicum, C. aurantibutylicum, C. beijerinckii* and *C.tetanomorphum* [69-71] *C.saccharoperbutylacetonicum, C.saccharoacetobutylicum, C.aurantibutyricum, C.pasyeurianum, C.sporogenes* and *C.Cadaveris* [72]. Current yields of bio-butanol remain low with wild type *C. acetobutylicum* rarely producing more than 13g/L, while chemically or genetically mutated strains of Clostridium rarely produce more than 20g/L *73, 74.* Research continues to improve butanol production by *Clostridium* through the use of metabolic and genetic engineering.

Hydrogen has also been the recent focus as a potential alternative fuel source because of its high energy yield (highest energy weight content per unit weight of all naturally occurring fuels [75], as well as its environmentally friendly nature due to its sole by-product of combustion being water [76]. *Clostridium* has shown to be a highly effective hydrogen producer, many strains of which have been isolated and further studied for their hydrogen producing potential [77-81]. Clostridia are able to utilize organic waste water and biomass to produce hydrogen [58].

Since the genome of Clostridia became accessible, increasing amounts of desired fermentation products, such as butanol, has become possible. The major principle of designing and improving clostridial strains is to increase amounts of the desired product is to decrease formation of other by products [62], for example, strategies to increase butanol production in *C. acetobutylicum* and recently included disrupting butyrate, acetone, lactate and acetate pathways [82], promoter exchange for overexpression of genes such as adhE1 (gene encoding acetylCoA acetytansferase) and down-regulation of other genes such as ctfA [74] as well as down regulation of acetone pathway [83] and co-production of riboflavin [84].

The ClosTron System and Metabolic Engineering-Emerging Biotechnology

TargeTron, a gene knockout system enabling disruption of genes in prokaryotic organisms [85], has been adapted for use for *Clostridium* named the ClosTron [86]. The ClosTron is a Group II intron directed mutagenesis system for Clostridium, has played a significant role in genetic manipulation of *Clostridium* [87]. Before the development of the ClosTron system, development of *Clostridium* mutants was mostly achieved by single cross-over knockouts by homologous recombination of a replication-defective plasmid [86]. This however, produced unstable mutants [87], a problem in which the ClosTron system overcame.

The ClosTron system allows directed insertional inactivation of genes using a bacterial group II intron, producing stable mutants [86, 87]. Stability in ensured by two key properties of the ClosTron system, including intron target specificity and ensuring intron mobility through an intron encoded protein (IEP) provided during mutagenesis then removed to ensure stability. Further modifications to improve the ClosTron system have been developed including the use of marker-recycling to develop strains with multiple mutations [87].

The ClosTron system has proved to be a vital application in genetic and metabolic engineering of *Clostridium*. For example, modifications of the ClosTron system to allow multiple knockouts mutants via flippase-mediated marker recycling provided an important method for sustainable metabolic engineering for Clostridial butanol production [87].

The Clostron system has also been used in the increased production of Ethanol in *C. acetobutylicum*through disruption of the butyrate/butanol fermentative pathway [88] as well as increase the production of hydrogen by Clostridia [89].

The Uses of Clostridia in Medical Contexts

Various Clostridium species have been utilised for their potential in various clinical applications including the use of Botox for numerous disorders, as drug delivery systems, in food poisoning prevention and in tumour therapy. With the continued advances in genetic engineering, further advances in these fields can be achieved.

Botox - Cosmetic and Medical Applications

Botulinum type A (BTX-A), one of seven neuromuscular blocking agents produced by *C. botulinum* [90], has been utilised in medicine for the treatment of a number of conditions and use in cosmetic enhancement, due to its ability to exert a paralytic effect on muscle [22].Table 3 below summarises the present clinical uses of BTX-A.

The toxin results in paralysis by inducing reversible cholinergic at the neuromuscular junction through four sequential steps [120]. Firstly BTX-A selectively binds to presynaptic cholinergic nerve terminals by its heavy chain. It is then internalised via receptor-mediated endocytosis [121].

From here, the light chain is released into the cytoplasm where it cleaves the synaptosome-associated protein (SNAP-25), that is necessary for neurotransmitter (acetocholine) release [90]. BTX-A was therefore designed to treat disorders with hyperfunction of cholinergic terminals such as spasticity and dystonia [122].

Due to limited action time of BTX-A, treatments are readministered every 3 to 4 months to remain affective [123]. Studies continue to determine if BTX-A can be used for treatment of other disorders such as cystitis or painful bladder syndrome [124].

Table 3. Clinical uses of BTX-A

Condition Type	Condition	Reference
Neurological	Blepharospasm	[91]
	Spasticity	[92]
	Dystonias	[93]
	Migraine	[94]
	Tourette's Syndrome	[95]
	Parkinson's disease	[96]
	Whiplash	[97]
Gastrological	Achalasia	[98]
	Sphincter of Oddi Dysfunction	[99]
	Anal Fissure	[100]
	Pelvic floor dysfunction	[101]
Reproductive	Cervical stenosis	[102]
	Idiopathic dyspareunia	[102]
	vaginismus	[102]
Dermatological	Hyperhidrosis	[103]
	Facial Flushing	[104]
	Pachydermoperiostosis	[105]
	Plantar fasciitis	[106]
Cosmetic Enhancement	Facial rejuvenation	[107]
Ophthalmology	Strabismus	[108]
Urology	Detrusor instability	[109]
	Detrusor-Sphincter dyssynergia	[110]
	Chronic prostate pain	[111]
Oral and Maxillofacial	Bruxism	[112]
	Masseteric hypertrophy	[113]
	Oromandibular dystonia	[114]
	Orofacial motortics	[115]
	Myofascial pain disorder	[116]
Salivary Gland	Sialorrhea	[117]
	Salivary gland fistulas	[118]
	sialoceles	[119]

Targeted Drug Delivery Systems

Clostridial neurotoxins for use as vehicles for drug delivery to neuronal cells haveshown great potential. Due to its capability of selectively targeting neuronal cells, the use of tetanus neurotoxin (TeNT), has been extensively studied for potential use as a drug delivery vector to the CNS [125]. Research has shown that when BIIb fragment of TeNT (fragment involved in recognition by nerve cell endings), is conjugated with glucose oxidase, the resulting conjugates are taken up by axon terminals and conveyed to the spinal cord motor neurons after intramuscular injection in mouse models [126].

Further research has shown improved delivery and bio-availability of cargo when human Cu/Zn superoxide dismutase was conjugated with the non-toxic TeNT C fragment [127], while promotion of motorneuron survival was observed when Cardiotrophin-1 was coupled with TeNT fragment C [128]. Delivery of Glial cell line neuro-trophic factor to the spinal cord was also shown to be improved when conjugated with TeNT fragment C protein [129].

These examples indicate a potential for TeNT to be used as a non-viral vehicle for treating neuronal conditions such as ALS, Alzheimer's, and spinal cord injury by providing a minimally invasive method of delivering drugs to multiple affected regions of the CNS. However, general immunity of the human population to Tetnus by vaccination with the tetanus toxoid, TeNT may not be an effective tool in human therapy, as antibodies would potentially interfere with the delivery vehicle [125].

Botulinium neurotoxin (BoNT),a clostridial neurotoxin originating from the species *C.botulinum*, may provide an alternative to using TeNT. BoNT shares similar modular structure, neuronal binding principles and belong to the same class of proteins as TeNT and are capable of delivering therapeutic cargos [125]. There is also far less concern of pre-existing immunity in comparison with TeNt.

Prevention of Food Poisoning

Nontoxic molecules of *C.perfringens* type A enterotoxin (CPE) have been proposed for use in prevention and treatment of food poisoning. These toxoids irreversibly bind to and saturate receptor sites on intestinal membranes therefore competing for receptor sites with CPE. *C.perfringens* toxoids are produced by inserting the enterotoxin gene fragment encoding CPE binding domain into a DNA vector, which is then inserted into *Escherichia coli. E.coli* then produces the *C.perfringens* toxoids from the recombinant DNA [130].

Targeted Cancer Treatment

Toxin as Therapeutic Agents

Toxins from particular *Clostridium* species have also been studied for their potential for targeted cancer treatment. CPE, associated with food poisoning and antibiotic- associated diarrhea [131], has been found to naturally bind to claudin molecules (CLDN)-4 as its high affinity receptor and CLDN-3 as its low affinity receptor [132] and are the only known CLDN proteins capable of mediating CPE binding and rapid cytolysis [133]. CPE bind to CLDN-4 through the C terminal 30 amino acids.

While expressed normally in intestinal epithelial cells, as well as the prostate and lungs at lower levels [134], CLDN-4 has been found to be over-expressed on the cell plasma membrane in uterine, ovarian, breast, prostate, heptocellular, colon, gastric and pancreatic cancers [135-141].

It was also found that CLDN-4 is not only related to metastasis, platinum resistance and poor survival in ovarian cancer [142], but CLDN-4 overexpression also promotes invasion, survival and motility of ovarian cancer cells associated with up-regulation of metalloproteinase-2 activity [143].

With this knowledge, we [144], constricted cDNA -comprised of C-CPE (C-terminus fragment of CPE sufficient for binding to target cell receptor but unable to promote large complex formation and cytolysis) and the toxic domain of *Pseudomonas aeruginosa* exotoxin A (ETA).ETAinhibits ADP-ribosylation of elongation factor 2 thereby inhibiting protein synthesis causing cell death [145]. Studies found that when fused with the variable domain of single chain antibody fragment (425scFv) that is specific for epidermal growth factor receptor (EGFR), the recombinant cytotoxic agent had a therapeutic effect on blood disorders and pancreatic cancer [146].

It was found that the recombinant C-CPE-ETA fusion protein specifically bound CLDN-4 and initiated penetration into cytosol in 5 cancers positive for CLDN-4 including breast, skin, colon and prostate cancers, but did not in CLDN-4 negative cells. It was also shown C-CPE-ETA was strongly cytotoxic to CLDN-4 positive cancer cells while not in CLDN-4 negative cancers.

This study further more demonstrated that three repeated injections of C-CPE-ETA reduced or completely abolished subcutaneously implanted MCF7 and SW480 xenograft tumours, therefore indicating the anti-cancer abilities of the recombinant fusion protein in CLDN-4 positive cancer models *in vivo* [144].

Live Clostridia Spores as Therapeutic Agents

When solid tumours grow too large for existing vasculatures in the surrounding normal tissue to support the tumour mass [147], some tumour cells become deficient in nutrients and oxygen and begin to accumulate waste products. This results in the release of tumour angiogenic factors which then stimulate endothelial cells of blood vessels in surrounding normal tissues to divide and migrate toward the tumour mass, and eventually form a network of new vessels within the tumour [147-149]. Although new blood can flow within the tumour mass, the architecture of the vascular network is chaotic and aberrant, and results in diminished oxygen delivery to the tumour cells causing a low oxygen microenvironment [149] as well as acidity, nutrient deficiency and cell death [150]. This characteristic of solid tumours provides an ideal growth environment for anaerobic bacteria such as *Clostridium*.

A major obstacle encountered in tumour treatments such as chemotherapy and radiotherapy, is the isolation of the treatment to the tumour. Studies have shown a potential for the use of clostridial spores to overcome this obstacle as when intravenously injected, clostridial spores are able to localise and germinate in the hypoxic or necrotic tissues commonly associated with solid tumours [151]. While Clostridia was first noted to cause tumour regression nearly two centuries ago when patients with gas gangrene begun to show tumour regression [152], it was in 1942, when it was first shown experimentally that injection of*C.histolyticum* spores into mice sarcomas resulted in lysis of tumour tissues [153]. Experiments following this discovery involved injection of *C.tetani* spore suspensions into mice with hepatomas and carcinomas [154]. While the control, tumour-free mice remained unaffected, the tumorous mice died within 48hours. Examination of the mice showed clostridial cells were isolated in the tumour [154].

These experiments established the potential for clostridial spores to germinate specifically in tumours following intravenous injection, prompting further analysis to determine if non-pathogenic strains could be distributed to a tumour with the intention of destroying the tumour/reducing its size through lysis [151]. Mose and Mose showed *C. butycum* strain M-55 could cause lysis in Ehrilich carcinomas of mice when spores localised and germinated within the tumour [155]. When similar results were achieved in subsequent experiments [156-158], Mose and Mose studied the effects of M-55 spores in humans with neoplastic disease after testing M-55 pathogenicity in humans. Between 10^9 and 10^{10} spores were injected into patients (intravenously or intra-tumourally). Variable tumour lysis was observed in all patients [159], while larger tumours underwent extensive lysis, only an outer rim of viable tumour tissue remained [160].

These studies showed a number of disadvantages including toxaemia, extensive growth of the Clostridia resulting in surgical intervention, regrowth from viable outer tumour tissue [159, 160]. In attempt to completely eradicate tumours, spore treatment was combined with other treatments including 5-fluordeoxyuridine, oxygen starvation, hyperthermia and x-rays along with many others [157, 158, 161-165]. These combinations however, did not succeed in completing tumour eradications and due to the regrowth of tumours, and therefore lack of clinical advantages,human clinical trials were ceased [166].

Advances in genetic engineering however, renewed interest in using Clostridia spores in tumour therapy.Brown introduced the concept to use non-pathogenic clostridia to deliver therapeutic proteins to tumours. Brown genetically engineered non-pathogenic *C.sporogenes* to produce specific anti-tumour proteins [151]. Once injected, spores would germinate only in hypoxic regions, such as those in tumour masses, therefore only distributing therapeutic proteins to the tumour [166]. Several studies showed varying levels of success when using Clostridia with anti-tumour proteins. One study showed that repeated treatment cycles were not only safe but treatment retained its efficacy. This study assessed the effects of multiple, consecutive treatment cycles, using a genetically modified *C. sporogenes,* and resulted in sustained tumour growth delay [167].

Successful preclinical results have been obtained using a new Clostridia strain, *C.novyi-NT* in combination with chemotherapy and radiation therapy [168, 169].

Control of tumour growth was also showed using *C.novyi-NT* and a single dose of liposomal doxorubicin [170]. For undisclosed reasons, Phase 1 safety study of *C.noyi-NT* spores in patients with treatment refractory solid tumour malignancies was terminated in July 2006 [171], however, the safety study was reopened in April 2011. The purpose of this study is to examine the safety of administration of *C. novyi-NT* spore in patients. It will also investigate the safety of using a sepsis scoring system to determine when antibiotics are required in these patients. Study primary completion date is set in December 2014 [172].

Conclusion

As this chapter has demonstrated, *Clostridium*has, over many decades, been utilized for a variety of purposes to benefit man-kind. Not only has this genus been of use in many industrial endeavours such as creation of environmentally chemicals, food production and processing and renewable biomass. It has also been of great benefit to the medical field, already helping to treat many conditions and hasshown potential in cancer treatment and drug delivery systems. With the continued advancement of genetic engineering, *Clostridium* will continue to demonstrate its potential in a variety ofapplicationsin both the industrial and medical fields, making this genus a vital part of our daily life and ourexciting future.

Acknowledgments

We thank other members of the Wei Laboratory for their helpful comments. This work was supported by the Dr. Jian Zhou smart state fellowship from Queensland government, and grants from the National Health and Medical Research Council and Cancer council, Queensland to MQW.

References

[1] Stang D. Clostridium clostridioforme (2011) Available From: http://zipcodezoo.com/Bacteria/C/Clostridium_clostridioforme/

[2] Johnson E. (2009) Clostridia. In: Schaechter M. *Encyclopedia of Microbiology (third edition)*. Academic Press. USA. 87-93.

[3] Cano RJ, Borucki MK. (1995) Revival and identification of bacterial spores in 25 to 40 million year old Dominican amber. *Science.* 268 (5213); 1060-1064.

[4] Vreeland RH, RosenweigWD, Powers DW. (2000) Isolation of a 250 million-year-old halotolerant bacterium from a primary salt crystal. *Nature*. 407(6806) 897-900.

[5] Xiao Y, Francke C, Abee T, Wells-Bennik. (2011) Clostridial spore germination versus bacilli: Genome mining and current insights. *Food Microbiology*.28(266-274).

[6] Popoff M B. (2009) Clostridial toxins. *Future Microbiology*. 4(8); 1021-1064.

[7] Wells, CL. Wilkins TD. (1996) Clostridia:Sporeforming Anaerobic bacilli. In: Baron S. *Medical Microbiology* (4th edition, c18).Galveston, Texas. University of Texas Medical Branch.

[8] Janvilisn T, Scaria J, Gleed R, Fubini S, Bonkosky M, Grohn Y, Chang. (2010) Development of a microarray for identification of pathogenic Clostridium spp. *Diagnostic Microbiol* 66; 140-147. Science Direct (www.science direct.com).

[9] Shimizu T, Ohtani K, Hirakawa H, Ohshima K, Yamashita A, Shiba T, Ogasawara N, Hattori M, Kuhara S, Hayashi H. (2002) Complete genome sequence of Clostridium perfringens, an anaerobic flesh-eater. *PNAS*. 99(2); 996-1001.

[10] Nolling J, Breton G, Omelchenko M, Makarova K, Zeng Q, Gibson R, Lee H, Qui D, Hitti J, Wolf Y, Tatusov R, Sabathe S, Doucette-Stamm L, Soucaille P, Daly M, Bennett G, Koonin E, Smith D. (2001) Genome sequence and comparative analysis of the solvent-producing bacterium Clostridium acetobutylicum. *J Bacteriol* 183(16): 4823-4828.

[11] Bruggemann H, Baumer S, Fricke W, Wiezer A, Liesegang H, Decker I, Herzberg C, Martinez-Arias R, Merkl R, Henne A, Gottschalk. (2002) The genome sequence of Clostridium tetani the causative agent of tetanus disease.*PNAS*. 100(3):1316-1321.

[12] Bettegowda C, Huang X, Lin J, Cheong I, Kohli M, Szabo S, Zhang X, Diaz L, Velculescu V, Parmigiani G, Kinzler K, Vogelstein B, Zhou S (2006) The genome and transcriptomes of the ant0-tumor agent Clostridium novyi-NT. *Nature biotechnology*. 24(12) 1573-1580.

[13] Sebaihai M, Wren B, Mullany P, Fairweather N, Minton N, Stabler R, Thomson N, Roberts A , Cerden˜o-Ta´rraga A, Wang H, Holden M, Wright A, Churcher C, Quail M, Baker S, Bason N, Brooks K, Chillingworth T, Cronin A, Davis P, Dowd L, Fraser A, Feltwell T, Hance Z, Holroyd S, Jagels K, Moule S, Mungall K, Price C, Rabbinowitsch E, Sharp S, Simmonds M, Stevens K, Unwin L, Whithead S, Dupuy B, Dougan G, Barrell B and Parkhill J. (2006). The Multidrug-resistant human pathogen Clostridium difficile has a highly mobile, mosaic genome. *Nature genetics*. 38(7). 779-786.

[14] Sebaihai M,Peck M, Minton N, Thomson N, Holden M, Mitchell W, Carter A, Bentley S, Mason D, Crossman L, Paul C, Ivens A, Wells-Bennik M, Davis I, Cerdeño-Tárraga A, Churcher C,Quail M, Chillingworth T, Feltwell T, Fraser A, Goodhead I, Hance Z, Jagels K, Larke N, Maddison M, Moule S, Mungall K, Norbertczak H, Rabbinowitsch E, Sanders M, Simmonds M, White B, Whithead S, and Parkhill J. (2007) Genome sequence of a proteolytic (Group I) *Clostridium botulinum* strain Hall A and comparative analysis of the clostridial genomes. *Genome research*. 17:1082-1092.

[15] Chen Y; Indurthi D, Jones S, Papoutsakis E. (2011) Small RNAs in the Genus Clostridium. *ClostridiumMicrobiology*. 2(1); 1-11.

[16] Waters L and Storz G. (2009). Regulatory RNAs in bacteria. *Cell*136:615– 628.

[17] Gottesman S. (2004). The small RNA regulators of *Escherichia coli*: roles and mechanisms. *Annu. Rev. Microbiol*. 58:303–328.

[18] Stackebrandt E, Hippe H (2001) Taxonomy and systematics. In: Bahl H, Dürre P. *Clostridia*. Weinheim, Germany; Willey-VCH; 19–48.

[19] Hatheway CL (1990) Toxigenic clostridia. *Clinical Microbiology. Rev*. 3:67-98.

[20] Stevens DL, Musher DM, Watson DA, Eddy H, Hamill RJ, Gyorkey F, Rosen H, Mader J.(1990) Spontaneous, nontraumatic gangrene due to *Clostridium septicum. Rev. Infect. Dis.* 12 (2); 286–296.

[21] Hauer P, Yeary T, Rosenbusch R (2004) Cloning and molecular characterization of the beta toxin (phospholipase C) gene of Clostridium haemolyticum. *Anaerobe* 10:243-254.

[22] Brin M.F (1997) Botulinum toxin: chemistry, pharmacology, toxicity and immunology. *Muscle and Nerve Supplement* 6, 146–168.

[23] Van Heyingen W.E.(1968). Tetanus. *Sci. Am*. 218; 69-77.

[24] Krug EL and Kent, C. (1984). Phospholipase C from Clostridium perfringens: preparation and characterisation of homogeneous enzyme. *Arch. Biochem. Biophys*. 231; 400–410.

[25] McClane BA (1996) An overview of Clostridium perfringens enterotoxin, *Toxicon* 34; 1335–1343.

[26] Sakurai J and Duncan C (1977) Purification of beta-toxin from Clostridium perfringens type C. *Infectious Immunology*. 18; 741-745.

[27] Uzal FA, Songer JG (2008) Diagnosis of *Clostridium perfringens* intestinal infections in sheep andgoats. *J. Vet. Diagn. Invest.* 20, 253–265.

[28] Goldstein J, Morris WE, Loidl CF, Tironi-Farinatti C, McClane BA, Uzal FA, Miyakawa ME. (2009) Clostridium perfringens epsilon toxin increases the small intestinal permeability in mice and rats. *PLoS One* 4(9); 7065.

[29] Ivie S, Fennessey C, Sheng J, Rubin D, McClain M. (2011) Gene-trap mutagenesis identifies mammalian genes contributing to intoxication by Clostridium perfringens ε-toxin. *Plos one*. 6(3) 1-13.

[30] Carman R, Perelle S, Popoff M. (1997). Binary Toxins from *Clostridium spiroforme* and *Clostridium perfringens*. IN Rood J, McClane B, Songer J and Titball R. The Clostridia:Molecular Biology and Pathogenesis. Academic press. London, UK 359-368.

[31] Sobel J. (2005) Botulism. *Clinical Infectious Disease.* 41, 1167-1173.

[32] Merson M, Dowell R. (1973) Epidemiologic clinical and laboratory aspects of wound botulism. *New England Journal of Medicine.*289, 1005-1010.

[33] Rings D. (2004) Clostridial disease associated with neurologic signs: tetnus, botulism and enterotoxemia. *Vetinary ClinicalNorth Am. Food and animal Practices.* 20; 379-391, vii-viii.

[34] Eklund M, Dowell J. (1987) Avian Botulism. Charles C Thomas, Springfield, IL, USA.

[35] Galey F. (2001) Botulism in the horse. *Vet.Clin. North As. Bovine Practices.* 17; 579-588.

[36] Songer J. (2005) Clostridial Diseases in domestic animals. In: Handbook on Clostridia. Durre P.

[37] Hookman P, Barkin JS. (2009) *Clostridium difficile* associated infection, diarrhea and colitis. *World J. Gastroenterol.* 15; 1554–1580.

[38] Oldfield EC. (2006) *Clostridium difficile* associated diarrhea: resurgence with a vengeance. *Rev. Gastroenterol. Disord.* 6; 79–96.

[39] Gibert M, Jolivet-Reynaud C, Popoff MR. (1997) b2 toxin, a novel toxin produced by *Clostridium perfringens. Gene* 203; 65–73.

[40] Keyburn AL, Boyce JD, Vaz P, Bannam TL, Ford ME, Parker D, Rubbo AD, Rood JI, Moore RJ.(2008) NetB, a new toxin that is associated with avian necrotic enteritis caused by *Clostridium perfringens.PLoS Pathog.* 4, e26 (2008).

[41] Lewis CJ, Naylor RD. (1998) Sudden death in sheep associated with *Clostridium sordellii. Vet. Rec.* 142; 417–421.

[42] Al-Mashat RR, Taylor DJ. (1983) *Clostridium sordellii* in enteritis in an adult sheep. *Vet. Rec.* 112, 19.

[43] Borriello SP, Carman RJ. (1983) Association of i-like toxin and *Clostridium* spiroforme with both spontaneous and antibiotic-associated diarrhea and colitis in rabbits. *J. Clin. Microbiol.* 17; 414–418.

[44] Rushdy AA, White JM, Ramsay ME. (2003) Tetanus in England and Wales, 1984–2000. *Epidemiol.Infect.* 130, 71–77.

[45] Gibson K, Bonaventure Uwineza J, Kiviri W, Parlow J. (2009) Tetanus in developing countries: a case series and review. *Can. J. Anaesth.* 56; 307–315.

[46] Durre P. (2001)From Pandora's Box to CornucopiaClostridia- A historical perspective. In: Bahl H, Durre P. *Clostridia: Biotechnology and Medical applications*. Weinheim Germany. Wiley; 1-18.

[47] Khalid A, Arshad M, Anjum M, Mahmood T, Dawson L. (2011) The anaerobic digestion of solid organic waste. *Waste management.* 31; 1937-1744.

[48] Lee, J., Song, J., Hwang, S. (2009). Effects of acid pre-treatment on bio hydrogen production and microbial communities during dark fermentation. *Bioresour. Technol.* 100; 1491–1493.

[49] Fantozzi F, Buratti C. (2009) Biogas production from different substrates in an experimental Continuously Stirred Tank Reactor anaerobic digester. *Bioresour. Technol.* 100; 5783–5789.

[50] Themelis NJ, Ulloa PA. (2007) Methane generation in landfills. *Renew. Energy* 32;1243–1257.

[51] Lastella G., Testa C., Cornacchia G., Notornicola M., Voltasio F., Sharma V.K. (2002). Anaerbic digestion of semi-solid organic waste: biogas production and its purification. *Energy Conserv. Manage.* 43; 63–75.

[52] Jingura, R.M., Matengaifa, R. (2009) Optimization of biogas production by anaerobic digestion for sustainable energy development in Zimbabwe. *Renew. Sust. Energy Rev.* 13; 1116–1120.

[53] Atsumi, S., Hanai, T., Liao, J.C. (2008) Non-fermentative pathways for synthesis of branched-chain higher alcohols as biofuels. *Nature* 451; 86–89.

[54] Zeng A.P., Biebl H. (2002) Bulk chemicals from biotechnology: the case of 1,3- propanediol production and the new trends. *Adv. Biochem. Eng. Biotechnol.* 74; 239–259.

[55] Zhang Y.,LiY.,DuC.,Liu,M.Cao,Z.(2006) Inactivation of aldehyde dehydrogenase: a key factor for engineering 1,3-propanediol production by Klebsiella pneumoniae. *Metab.Eng.* 8;578–586.

[56] Kalscheuer R.,StoltingT.,SteinbuchelA.(2006) Microdiesel: Escherichia coli engineered for fuel production. *Microbiology-SGM* 152;2529–2536.

[57] Schirmer, A., Rude, M.A., Li, X., Popova, E., del Cardayre, S.B. (2010) Microbial biosynthesis of alkanes. *Science* 329; 559–562.

[58] Zhao X, Xing D, Fu N, Liu B, Ren N. (2011) Hydrogen Production by the newly isolated Clostridium beijerinckii RZF-1108. *Bioresource Technology*. 1-5: doi:10.1016/j.biortech.2011.02.086.

[59] Tran H, Cheirsilp B, Hodgson B, Umsakul K. (2010) Potential use of Bacillus subtilise in a co-culture with Clostridium butylicum for acetone-butanol-ethanol production from cassava starch. *Biochemical Engineering Journal*. 48: 260-267.

[60] Jones, TJ.,woods D.T. Jones, D.R. Woods. (1986) Acetone–butanol fermentation revisited, *Microbiol. Rev*.50 (1986) 484–524.

[61] Gheshlaghi R, Scharer J, Moo-Young M, Chou C. (2009) Metabolic pathways of clostridia for producing butanol. *Biotechnology Advances*. 27:764-781.

[62] Lutke-Eversloh T, Bahl H. (2011) Metabolic engineering of Clostridium acetobutylicum: recent advances to improve butanol production. *Current opinon in Biotechnology*. 22: 1-14.

[63] Durre P. (2007) Biobutanol: an attractive biofuel. *Biotechnol J*, 2:1525-1534.

[64] Rose AH. (1961) Industrial microbiology. UK, London: Butterworths.

[65] Zhu L, Dong H, Zhang Y, Li. (2011) Engineering the robustness of Clostridium acetobutylicum by introducing glutathione biosynthetic capability. *Metabolic Engineering*. doi:10.1016/j.ymben.2011.01.009.

[66] Huang H, Liu H, Gan Y. (2010) Genetic modification of critical enzymes and involved genes in butanol biosynthesis from biomass. *Biotechnology Advances*. 28:651-657.

[67] Montoya D, Arevalo C, Gonzales S, Aristizabal F, Schwarz W. (2001) New Solvent-producing Clostridium sp. Strains, hydrolysing a wide range of polysaccharides, are closely related to Clostridium butyricum. *Journal of Industrial Microbiology and Biotechnology*.27(5); 329-335.

[68] Jones TJ, Woods DR. Solvent production. In: Minton NP, Clarke DJ. *Clostridia*. USA, NY: Plenum Press; 1989. p. 105–44.

[69] Gottwald M, Hippe H, Gottschalk G.(1984) Formation of n-butanol from D-glucose by strains of "Clostridium tetanomorphum" group. *Appl Environ Microbiol*; 48:573–6.

[70] George HA, Johnson JL, Moore WEC, Holdeman LV, Chen JS. (1983) Acetone, isopropanol, andbutanol pr oduction by Clostridium beijerinckii (syn. Clostridium butylicum) and Clostridium aurantibutylicum. *Appl Environ Microbiol* 45:1160–3.

[71] George HA, Chen JS. (1983) Acidic conditions are not obligatory for onset of butanol formation by *Clostridium beijerinckii* (synonym. *C. butylicum*). *Appl Environ Microbiol* ;46: 321–7.

[72] Keis S, Bennett CF, Ward VK, Jones DT (1995) Taxonomy and phylogeny of industrial solvent-producing clostridia. *Int. J. Syst. Bacteriol.* 45:693– 705.

[73] Ezeji, T., Milne, C., Price, N.D., Blaschek, H.P.(2010) Achievements and perspectives to overcome the poor solvent resistance in acetone and butanol-producing microorganisms. *Appl. Microbiol. Biotechnol.* 85; 1697–1712.

[74] Sillers R, Al-Hinai MA, Papoutsakis ET. (2008) Aldehyde-alcohol dehydrogenase and/or thiolase overexpression coupled with CoA transferase down regulation lead to higher alcohol titers and selectivity in Clostridium acetobutylicum fermentations. *Biotechnol Bioeng*, 102:38-49.

[75] Das, D. (2009) Advances in biohydrogen production processes: an approach towards commercialization. *Int. J. Hydrogen Energy* 34, 7349–7357.

[76] Zhang Y, Shen J.(2007) Enhancement effect of gold nanoparticles on biohydrogen production from artificial wastewater. *Int J Hydrogen Energy*; 32:17–23.

[77] Jo J.H., Jeon C.O., Lee S.Y., Lee D.S., Park J.M. (2010) Molecular characterization and homologous overexpression of [FeFe]-hydrogenase in Clostridium tyrobutyricum JM1. *Int. J. Hydrogen Energy* 35, 1065–1073.

[78] Lin C.Y., Shei S.H. (2008) Heavy metal effects on fermentative hydrogen production using natural mixed microflora. *Int. J. Hydrogen Energy* 33, 587–593.

[79] Morimoto K., Kimura T., Sakka K., Ohmiya K. (2005) Overexpression of a hydrogenase gene in Clostridium paraputricum to enhance hydrogen gas production. *FEMS Microbiol. Let.* 246; 229–234.

[80] Wang M.Y., Olson B.H., Chang J.S. (2008). Relationship among growth parameters for Clostridium butyricum, hydA gene expression, and biohydrogen production in a sucrose-supplemented batch reactor. *Appl. Microbiol. Biotechnol.* 78, 525– 532.

[81] Wang J., Wan W. (2008). Effect of Fe2+ concentration on fermentative hydrogen production by mixed cultures. *Int. J. Hydrogen Energy* 33, 1215–1220.

[82] SoucailleP. (2010) Process for the biological production of n-butanol at high yield. International patent WO2008/052973.

[83] Jiang Y,XuC,DongF,YangY,JiangW,YangS. (2009) Disruption ofthe acetoacetate decarboxylase gene in solvent-producing Clostridium acetobutylicum increases the butane ratio. *Metab. Eng*, 11:284-291.

[84] Cai X,Bennett B.(2010) GN: Improving the Clostridium acetobutylicum butanol fermentation by engineering the strain for co- production of riboflavin. *J. IndMicrobiol. Biotechnol.* doi: 10.1007/s10295-010-0875-6.

[85] Sigma-Aldrich Co. (2011) Targe Tron: Genetic engineering that is right on target. Available at: http://www.sigmaaldrich.com/life-science/functional-genomics-and-rnai/targetron.html

[86] Heap J, Pennington O, Cartman S, Carter G, Minton N. (2007) The ClosTron: A universal gene knockout system for the genus Clostridium. *Journal of Microbiological methods*. 70:452-464.

[87] Heap J, Kuehne S, Ehsaan M, Cartman S, Cooksley C, Scott J, and Minton N. (2010) The ClosTron: Mutagenesis in Clostridium refined and streamlined. *Journal of Microbiological methods*. 80:49-55.

[88] Lehmann D and Lutke-Eversloh T. (2011) Switching Clostridium acetobutylicum to an ethanol producer by disruption of the butyrate/butanol fermentative pathway. *Metabolic engineering*. doi:10.1016/j.ymben.2011.04.006.

[89] Hashesh M, Wang R, Hallenbeck P. (2011) Metabolic engineering in dark fermentative hydrogen production; theory and practice. *Bioresource Technology*. doi:10.1016/j.biortech.2011.03.016.

[90] Fortuna R, Vaz M, Youssef A, Longino D, Herzog W. (2011) Changes in contractile properties of muscles recieveing repeat injections of botulinum toxin (Botox). *Journal of Biomechanics*. 44:39-44.

[91] Wirtschafter JD, McLoon LK. (1998) Long-term efficacy of local doxorubicin chemomyectomy in patients with blepharospasm and hemifacial spasm. *Ophthalmology*;105(2):342e6.

[92] Papadonikolakis AS, Vekris MD, Korompilias AV, Kostas JP, Ristanis SE, Soucacos PN. (2003) Botulinum A toxin for treatment of lower limb spasticity in cerebral palsy: gait analysis in 49 patients. *Acta Orthop Scand*;74(6):749e55.

[93] Sutcliffe R, Sandiford N, Khawaja. (2005) From frown lines to fissure: Therapeutic uses for botulinum toxin. *International Journal of Surgery*. 3; 141-146.

[94] Ashkenazi A, Silberstein SD. (2004) Botulinum toxin and other new approaches to migraine therapy. Annu Rev Med; 55: 505-518.

[95] Porta M, Maggioni G, Ottaviani F, Schindler A. (2004) Treatment of phonic tics in patients with Tourette's syndrome using botulinum toxin type A. *Neurol Sci*; 24(6):420e3.

[96] Ondo WG, Hunter C, Moore W. (2004) A double-blind placebocontrolled trial of botulinum toxin B for sialorrhea in Parkinson's disease. *Neurology*; 62(1):37-40.

[97] Juan J. (2004). Use of botulinum toxin-A for musculoskeletal pain in patients with whiplash associated disorders. *BMC Musculoskelet Disord*; 5(1):5.

[98] Pasricha PJ, Ravich WJ, Hendrix TR, Sostre S, Jones B, Kalloo AN. (1995) Intrasphincteric botulinum toxin for the treatment of achalasia. *N Engl J Med*; 332(12):774e8.

[99] Sherman S, Ruffolo TA, Hawes RH, Lehman GA. (1991) Complications of endoscopic sphincterotomy. A prospective series with emphasis on the increased risk associated with sphincter of Oddi dysfunction and nondilated bile ducts. *Gastroenterology*; 101(4):1068-1075.

[100] Lindsey I, Jones OM, Cunningham C, Mortensen NJ. (2004) Chronic anal fissure. *Br J Surg*; 91(3):270-279.

[101] Maria G, Sganga G, Civello IM, Brisinda G. (2002) Botulinum neurotoxin and other treatments for fissure-in-ano and pelvic floor disorders. *Br J Surg* 2002;89(8):950-961.

[102] Berge C and Cronje H. (2011) Cervical stenosis relieved by botulinum toxin A injection. *Int J Gynecol Obst*. 112: 243-251.

[103] Bhidayasiri R, Truong DD. (2008) Evidence for effectiveness of botulinum toxin for hyperhidrosis. *J Neural Transm*; 115: 641-645.

[104] Yuraitis M, Jacob CI. (2004) Botulinum toxin for the treatment ofvfacial flushing. *Dermatol Surg*; 30(1):102-104.

[105] Ghosn S, Uthman I, Dahdah M, Kibbi AG, Rubeiz N. (2010) Treatment of pachydermoperiostosis pachydermia with botulinum toxin type A. *J. Am. Acad. Dermatol.* 63(6); 1063-1041

[106] Chou LW, Hong CZ, Wu ES, Hsueh WH, Kao MJ. (2011) Serial ultrasonographic findings of plantar fasciitis after treatment with botulinum toxin a: a case study.*Arch. Phys. Med. Rehabil;* 92(2):316-9.

[107] Carruthers JA, Lowe NJ, Menter MA, Gibson J, Nordquist M, Mordaunt J, Walker P, Eadie N. (2002) BOTOX; Glabellar Lines I Study Group. A multicenter, double-blind, randomized, placebo-controlled study of the efficacy and safety of botulinum toxin type A in the treatment of glabellar lines. *J. Am. Acad. Dermatol.*; 46(6):840-849.

[108] Scott AB, Magoon EH, McNeer KW, Stager DR. (1990) Botulinum treatment of childhood strabismus. *Ophthalmology*; 97(11):1434-1438.

[109] Dykstra D, Enriquez A, Valley M. (2003) Treatment of overactive bladder with botulinum toxin type B: a pilot study. *Int. Urogynecol. J. Pelvic Floor Dysfunct;* 14(6):424-426.

[110] Dykstra DD, Sidi AA. (1990) Treatment of detrusor-sphincter dyssynergia with botulinum A toxin: a double-blind study. *Arch Phys Med. Rehabil*; 71(1):24-26.

[111] Zermann D, Ishigooka M, Schubert J, Schmidt RA.(2000) Perisphincteric injection of botulinum toxin type A. A treatment option for patients with chronic prostatic pain? *Eur Urol* 38(4):393e9.

[112] Tan E, Jankovic J. (2000) Treating Severe bruxism with botulinum toxin. *Clin Pharmacol.* 38(4) 393-399.

[113] Smyth AG. (1993) Botulinum toxin treatment of bilateral massenteric hypertrophy. *British J Oral Maxxillofacial Surg*. 32(1); 29-33.

[114] Tan E, Jankovic J. (1999) Botulinum toxin A in patients with oromandibular dystonia. *Neurology*. 53(9) 2102.

[115] Evidente V, Adler C. (1998) Hemifacial spasm and other craniofacial movement disorders. *Mayo Clinic Proceeding.* 73 (1); 67-71.

[116] Porta M. (1999) A comparative trial of botulinum toxin type A and methylprednisolone for the treatment of myoascial pain syndrome and pain from chronic muscle spasm. *J Int Assoc study pain*. 85(1) 101-105.

[117] Benson J, Daugherty K. (2007) Botilinum Toxin A in the Treatment of Sialorrhea. *Ann Pharmacol.* 41(1); 79-85.

[118] Lim Y, Choi E. (2007) Treatment of an acute salivary fistula after parotid surgery: botulinum toxin type A injection as primary treatment. *European Archives of Oto-Rhino-Laryngology*. 265(2); 243-245.

[119] Vargas H, Galati LT, Parnes SM. (2000) A pilot study evaluating the treatment of postparotidectomy sialoceles with botulinum toxin type A. *Arch Otol Head Neck Surg*; 126:421–424.

[120] Wenzel RG. (2004) Pharmacology of botulinum neurotoxin serotype A. *Am J Health Syst Pharm*61(22 suppl 6):S5–S10.

[121] Black JD, Dolly JO. (1986) Interaction of 125I-labeled botulinum neurotoxins with nerve terminals. II. Autoradiographic evidence for its uptake into motor nerves by acceptormediated endocytosis. *J Cell Biol* 103(2):535e44.

[122] Lange, F., Weerden, T.W., Hoeven, J.H. (2007). Age-related botulinum toxin effects on muscle fiber conduction velocity in non-injected muscles. *Clin Neurophysiol* 118, 2398–2403.

[123] Inagi, K., Connor, N.P., Schultz, E.,Ford, C.N., Cook ,C.H., Heisey, D.N. (1999). Muscle fiber-type changes induced by botulinum toxin injection in the rat larynx. *Otolaryngology—Head NeckSurg*120,876–883.

[124] Gottsch H, Miller J, Yang C, Berger R. (2011) A Pilot study of botulinum for interstitial cystitis/painful bladder syndrome. *NeurolUrodynamics.* 30;93-96.

[125] Singh B, Thirunavukkarasu N, Ghosal K, Ravichandran E, Kukreja R, Cai S, Zhang P, Ray R, Ray P. (2010) Clostridial neurotoxins as a drug delivery vehicle targeting nervous system. *Biochimie.* 92:1252-1259.

[126] Beaude P, Delacour A, Bizzini B, Domuado D, Remy MH. (1990) Retrograde axonaltransport of an exogenous enzyme covalently linked to B-IIb fragment of tetanus toxin. *Biochem. J.* 271; 87-91.

[127] Benn SC, I. Ay, E. Bastia, R. Chian, S.A. Celia, R.B. Pepinsky, P.S. Fishman,R.H. Brown Jr., J.W. Francis (2005) Tetanus toxin fragment C fusion facilitates protein delivery to CNS neurons from cerebrospinal fluid in mice. *J. Neurochem.* 95, 1118e1131.

[128] Bordet T, Ptakhine LC, Fauchereau F, Friocourt G, Kahn A, Haase G. (2001) Neuronal targeting of cardiotrophin-1 by coupling with tetanus toxin C fragment. *Mol. Cell. Neurosci.* 17; 842-854.

[129] Larsen KE, Benn SC, Ay I, Chian RJ, Celia SA, Remington MP, Bejarano M, Liu M, Ross J, Carmillo P, Sah D, Phillips KA, Sulzer D, Pepinsky RB, Fishman PS, Brown RH., Francis JW. (2006) A glial cell line-derived neurotrophic factor (GDNF): tetanus toxin fragment C protein conjugate improves delivery of GDNF to spinal cord motor neurons in mice. *Brain Res.* 1120; 1-12.

[130] McClane B, Hanna P, Mietzner T (1997) Clostridium Perfringens Type A enterotoxin and methods of preparation and use as a vaccine and therapeutic agent. US Patent number: 5,695,956.

[131] McClane BA. (1996) An overview of *Clostridium perfringens* enterotoxin, *Toxicon* 34; 1335–1343.

[132] Tsukita s, Furuse M, Itoh M. (2001) Multifunctional strands in tight junctions, *Nat. Rev. Mol Cell Biol.* 2; 285–293.

[133] Katahira J, Sugiyama H, Inoue N, Horiguchi Y, Matsuda M, Sugimoto N. (1997) Clostridium perfringens enterotoxin utilizes two structurally related membrane proteins as functional receptors in vivo, *J. Biol. Chem.* 272; 26652– 26658.

[134] Kominsky SL, Vali M, Korz D, Gabig TG, Weitzman SA, Argani P, Sukumar S. (2004) Clostridium perfringens enterotoxin elicits rapid and specific cytolysis of breast carcinoma cells mediated through tight junction proteins claudin 3 and 4.*Am. J. Pathol.* 164; 1627–1633.

[135] Landers KA, Samaratunga H, Teng L, Buck M, Burger MJ, Scells B, Lavin MF, Gardiner RA. (2008) Identification of claudin-4 as a marker highly overexpressed in both primary and metastatic prostate cancer, *Br. J. Cancer.* 99; 491–501.

[136] Nichols LS, Ashfaq R, Iacobuzio-Donahue CA. (2004) Claudin 4 protein expression in primary and metastatic pancreatic cancer: support for use as a therapeutic target, *Am .J. Clin. Pathol.* 121; 226–230.

[137] Kulka J, Szasz AM, Nemeth Z, Madaras L, Schaff Z, Molnar IA, Tokes AM. (2009) Expression of tight junction protein claudin-4 in basal-like breast carcinomas, *Pathol. Oncol. Res.* 59–64.

[138] Litkouhi B, Kwong J, Lo CM, Smedley JG, McClane BA, Aponte M, Gao Z, Sarno JL, Hinners J, Welch WR, Berkowitz RS, Mok SC, Garner EI. (2007) Claudin-4 overexpression in epithelial ovarian cancer is associated with hypomethylation and is a potential target for modulation of tight junction barrier function using a C-terminal fragment of *Clostridium perfringens* enterotoxin.*Neoplasia* 9; 304–314.

[139] Lodi C, Szabo E, Holczbauer A, Batmunkh E, Szijarto A, Kupcsulik P, Kovalszky I, Paku S, Illyes G, Kiss A, Schaff Z. (2006) Claudin-4 differentiates biliary tract cancers from hepatocellular carcinomas, *Mod. Pathol.* 19; 460– 469.

[140] Semba S, Hasuo T, Satake S, Nakayama F, Yokozaki H. (2008) Prognostic significance of intestinal claudins in high-risk synchronous and metachronous multiple gastric epithelial neoplasias after initial endoscopic submucosal dissection, *Pathol. Int.* 58; 371–377.

[141] Halder SK, Rachakonda G, Deane NG, Datta PK. (2008) Smad7 induces hepatic metastasis in colorectal cancer, *Br. J. Cancer* 99; 957–965.

[142] Kleinberg l, Holth A, Trope CG, Reich R, Davidson B. (2008) Claudin upregulation in ovarian carcinoma effusions is associated with poor survival, *Hum. Pathol.* 39; 747–757.

[143] Agarwal R, D'Souza T, Morin PJ. (2005) Claudin-3 and claudin-4 expression in ovarian epithelial cells enhances invasion and is associated with increased matrix metalloproteinase-2 activity, *Cancer Res.* 65; 7378– 7385.

[144] Yao Q, Cao S, Li C, Mengesha A, Low P, Kong B, Dai S, Wei M. (2010) Turn a diarrhoea toxin into a receptor-mediated therapy for a plethora of CLDN-4-overexpressing cancers. *Biochem. Biophys. Res. Commun.*398(3); 413-419.

[145] Iglewski BH, Kabat D. (1975). NAD-dependent inhibition of protein synthesis by Pseudomonas aeruginosa toxin. *Proc.Natl. Acad. Sci.* USE 72:2284-2288.

[146] Bruell D, Stocker M, Huhn M, Redding N, Kupper M, Schumacher P, Paetz A, Bruns CJ, Haisma HJ, Fischer R, Finnern R, Barth. (2003) The recombinant anti-EGF receptor immunotoxin 425(scFv)-ETA' suppresses growth of a highly metastatic pancreatic carcinoma cell line. *In J Oncol* 23: 1179-1186.

[147] Folkman J.(1990) What is the evidence that tumours are angiogenesis dependent? *J Natl Cancer Inst*; 82:4–6.

[148] Folkman J. (1995) Angiogenesis in cancer, vascular, rheumatoid and other disease. *Nat Med*; 1:27–31.

[149] Konerding, M.A., Miodonski, A.J., Lametschwandtner, A. (1995). Microvascular corrosion casting in the study of tumor vascularity: a review. *Scanning Microsc.* 9, 1233–1244.

[150] Wei MQ, Ellem K, Dunn P, West M, Bai C, Vogelstein B. (2007) Facultative or obligate anaerobic bacteria have the potential for multimodality therapy of solid tumours. *Eur J cancer.* 43: 490-496.

[151] Minton NP, Mauchline ML, Lemmon MJ,Brehm JK, Fox M, Michael NP, Giaccia A, Bron JM. (1995)Chemotherapeutic tumour targeting using clostridial spores. *FEMS Microbiol Rev* 17(3): 357-64.

[152] Mengesha A, Dubois L, Chiu RK, *et al.* (2006)Potential and limitations of bacterial-mediated cancer therapy. *Front Biosci* 2007; 12: 3880-91.

[153] Parker, R.C., Plummber, H.C., Siebenmann, C.O. and Chapman, M.G. (1947) Effect of histolyticus infection and toxin on transplantable mouse tumours. *Proc. Soc. Exp. Biol. Med.* 66, 461-465.

[154] Malmgren, R.A. and Flanigan, C.C. (1955) Localisation of the vegetative form of *Clostridium tetani* in mouse tumours following intravenous spore administration. *Cancer Res.* 15, 473-478.

[155] Mose, J.R. and Mose, G. (1959) Onkolyseversuche mit apathogenen, anaeroben Sporenbildern am Ehrlich-Tumour der Maus. *Z. Krebsforsch.* 63, 447-455.

[156] Mose, J.R. and Mose, G. (1964) Oncolysis by clostridia. 1. Activity of *Clostridium butyricum* (M-55) and other non-pathogenic clostridia against the Ehrlich carcinoma. *Cancer Res.* 24, 212-216.

[157] Gericke, D and Engelbart, K. (1964) Oncolysis by clostridia. II. Experiments of a tumour spectrum with a variety of clostridia in combination with heavy metal. *Cancer Res.* 24, 217-221.

[158] Thiele, E.H., Arison, R.N. and Boxer G.E. (1964) Oncolysis by clostridia. IlL Effects of clostridia and chemotherapeutic agents on rodent tumours. *Cancer Res.* 24, 222-233.

[159] Heppner, F. and Mose, J.R. (1978) The liquefaction (oncolysis) of malignant gliomas by a nonpathogenic *Clostridium. Acta. Neurochir.* 42, 123-125.

[160] Carey, R.W., Holland, J.F., Whang, H.Y., Neter, E. And Bryant, B. (1967) Clostridial oncolysis in man. *Eur. J. Cancer* 3, 37-46.

[161] Schlechte, H., Schwabe, K., Mehnert, W.H., Schulze, B. And Br~iuniger, H. (1982) Chemotherapy for tumours using clostridial oncolysis, antibiotics and cyclophospamide: model trial on the UVT 15264 tumour. *Arch.Geschwulstforsch.* 52, 41-48.

[162] Kretschmer, H., Ludewig, R. and Hambsch, K. (1975) Treatment experiments with *Clostridium butyricum-spores* plus cyclophosphamide on the Jensen sarcoma of rats. *Arch.Geschwulstforsch.* 45, 16-18.

[163] Dietzel, F. and Gericke, D. (1977) Identification of the oncolysis by clostridia by means of radiofrequency hyperthermy in experiments on animals: dependence on dosage and intervals. *Strahlentherapie* 153, 263-266.

[164] Mose, J.R., Mose, M. and Fischer, G. (1980) Histamine reduction by means of oncolytically active and oncolytically non-active clostridial strains. *Zentralbl. Bakteriol. 1. Abt. Orig. A* 246, 541-549.

[165] Hambsch, K., Korner, CH., and Ludewig, R. (1982) Treatment experiments with glucose-induced hyperglycaemia plus *Clostridium-spores* on the Jensen sarcoma of rats. *Arch. Geschwulstforsch.* 52, 267-273.

[166] Mengesha A, Wei J, Zhou SF, Wei MQ. (2010) Clostridial spores to treat solid tumours- potential for a new therapeutic modality. *Current Gene Therapy.* 10: 15-26.

[167] Theys J, Pennington O, Dubois L, Anlezark G, Vaughan T, Mengesha A, Landuyt W, Anne J, Burke PJ, Curre P, Wouters BG, Minton NP, Lambin P. (2006)Repeated cycles of Clostridium- directed enzyme prodrug therapy result in sustained antitumour effects *in vivo*. *Br. J. Cancer*; 95(9): 1212 - 1219.

[168] Dang LH, Bettegowda C, Huso DL, Kinzler KW, Vogelstein B. (2001) Combination bacteriolytic therapy for the treatment of experimental tumors. *Proc. Natl. Acad. Sci. USA*; 98(26): 15155-60.

[169] Bettegowda C, Dang LH, Abrams R, Huso DL, Dillehay L, Cheong I, Agrawal N, Borzillary S, McCaffery JM, Watson EL, Lin KS, Bunz F, Baidoo K, Pomper MG, Kinzler KW, Vogelstein B, Zhou S. (2003)Overcoming the hypoxic barrier to radiation therapy with anaerobic bacteria. *Proc. Natl. Acad. Sci. USA* 100(25): 15083- 150838.

[170] Cheong I, Huang X, Bettegowda C, Diaz JA Jr, Kinzler KW, Zhou S, Vogelstein B. A (2006) Bacterial protein enhances the release and efficacy of liposomal cancer drugs. *Science* 314(5803): 1308-11.

[171] Diaz L. One Time Injection of Bacteria to treat solid tumors that have not responded to standard therapy. (December 2008). Available at: (http://clinicaltrialsfeeds.org/clinical-trials/show/NCT00358397.

[172] Rubin C. Clinical Trials Feed: Safety study of live bacteria to treat solid tumours that have not responded to standard therapy. (April 2011) Available at: http://clinicaltrialsfeeds.org/clinical-trials/show/NCT0111 8819.

In: Clostridia ISBN: 978-1-62100-761-6
Editors: S. V. Clinton and M. F. Kelly © 2012 Nova Science Publishers, Inc.

Chapter 3

CLINICAL ASPECTS OF *CLOSTRIDIUM DIFFICILE* ASSOCIATED DIARRHOEA

Helena A. White and Martin J. Wiselka
Department of Infection and Tropical Medicine,
Leicester Royal Infirmary, Infirmary Square, Leicester, UK

ABSTRACT

Clostridium difficile is an important healthcare associated infection, in terms of incidence, associated morbidity and mortality, and economic impact. The recent emergence of a more virulent strain, ribotype 027, has resulted in an increased incidence of *Clostridium difficile*- associated diarrhoea (CDAD) across the world, with an attendant rise in the severity of cases. This article will review the diagnosis, treatment, and control of the disease, and discuss new and emerging management strategies.

C. difficile was first isolated from neonatal bowel flora in 1935, and was subsequently linked to antibiotic related diarrhoea in the 1970s. CDAD commonly results from the use of antimicrobials in patients who often have underlying medical or surgical comorbidities, and causes profuse diarrhoea after disruption of the normal bowel ecoflora. The production of toxins A and B induces a pro-inflammatory response within the bowel, with epithelial cell apoptosis, and culminates in watery diarrhoea. The spectrum of disease ranges from self-limiting diarrhoea to severe colitis and toxic megacolon, and the infection frequently recurs in debilitated patients. Diagnosis of the condition depends on a high index of suspicion in vulnerable patients, with stool culture for *C. difficile* or more commonly, faecal toxin detection.

Treatment of CDAD is with antibiotics, most commonly metronidazole or vancomycin, and prolonged or gradually tapering courses are often needed in recalcitrant cases. Novel techniques such as faecal biotherapy, to restore the natural bowel ecosystem, are used in some centres. Close attention to nutrition is needed, with supplementary enteral feeding commonly employed in debilitated patients. Hospital inpatients often have a slow recovery and require enthusiastic multi-disciplinary management. Mortality rates vary, and are related both to the underlying health status of the host, and the severity of the clinical illness.

Persistence of spores in the environment presents a challenge to healthcare facilities in terms of preventing nosocomial acquisition, and re-infection of treated cases. Strategies such as antimicrobial stewardship, cohorting of affected patients, environmental cleansing and improved staff hand hygiene have all been shown to be of benefit in the control of this disease.

Introduction

Clostridium difficile is a fastidiously anaerobic, gram positive bacillus, which is both spore and toxin – producing. Its primary medical significance is that it causes profuse watery diarrhoea in affected patients, and is an important transmissible nosocomial pathogen, responsible for large institutional outbreaks of infection [1,2,3].

The organism was first isolated in 1935 from neonatal bowel flora [4] and thus named owing to the difficulty with which it was isolated and cultured on conventional media. Some thirty years later, the organism was linked to cases of pseudomembranous colitis. The endoscopic appearance of yellow, "diphtheric-like" membranes in the colon had been documented since 1893, and in the pre-antibiotic era, occurred following bowel surgery or ischaemia, and other severe systemic illnesses including sepsis and renal failure [5]. Following the introduction of antibiotics, pseudomembranous colitis was initially attributed to nosocomial acquisition of *Staphylococcus aureus*, but work in the 1970s identified *C. difficile* toxin in the stools of patients who had received clindamycin antibiotic therapy, and the pathogenic nature of the organism was realised [6].

The Importance of Infection

C. difficile is frequently found in the stools of young children under the age of two years, and in a few percent of healthy adults, where it represents asymptomatic colonisation. However, amongst hospital inpatients with medical and surgical comorbidities, it can be responsible for severe episodes of diarrhoea, with attendant mortality. It is the aetiological agent of up to 30% of the total cases of antibiotic-related diarrhoea, and overall is the most common cause of healthcare-associated infection. The predominant trigger for development of clinical infection is disruption of the normal bowel ecoflora in a colonised patient, usually occurring as a consequence of antibiotics administered for another infection [7].

The production of spores enables the organism to persist in the environment, and poses a major challenge to healthcare facilities during attempts to eradicate the infection. Spores are resistant to heat, antibiotics and acid, but after ingestion, they vegetate within the human colon to a functional, toxin-producing state, resulting in the symptoms of clinical infection. The organism produces two main toxins, Toxin A (tcdA) and Toxin B (tcdB).

Toxins A and B bind to the epithelial surface, and cause disruption of tight epithelial membranes, a secondary effect of which is the migration of inflammatory cells within the bowel lumen. Toxins additionally stimulate the production of cytokines, which, in conjunction with neutrophils, contribute to the formation of pseudomembranous plaques [8,9]. Epithelial cell apoptosis and the production of hydrolytic enzymes lead to a loss of integrity in the mucosal barrier, resulting in diarrhoea.

Since 2003, the emergence of a quinolone-resistant, hyper-virulent strain has been associated with nosocomial outbreaks across the U.S., Europe, and further afield including Australasia [2]. Terminology for this strain includes its restriction endonuclease analysis pattern, B1; its pulsed-field gel electrophoresis pattern NAP1 (North American pulsed-field type 1); and its PCR ribotype designation 027. NAP1/B1/027 possesses a deletion in the tcdC regulatory gene, which is situated downstream from, and controls the expression of toxins A (tcdA) and B (tcdB) [10]. This mutation enables the strain to sporulate more frequently and for longer than previously studied strains [11], which may contribute to its virulence and the increased relapses noted with its emergence. In addition to toxins A and B, this strain also encodes a gene for production of a binary toxin. This, along with toxins from different Clostridia species, has been demonstrated to induce the formation of a microtubule system which may increase bacterial adhesion to the bowel wall [12].

In the United States, incidence of CDAD- related deaths rose from 5.7 deaths per million population in 1999 to 23.7 deaths per million population in 2004 [13], as a result of the emergence of the epidemic strain. The Health Protection Agency (HPA) in England and Wales reported a substantial rise in the number of cases of infection, and subsequently made reporting of all cases of *C. difficile* infections in those over two years of age a mandatory obligation by hospital Trusts in April 2007 [14]. There has been a steady reduction in the number of cases as the epidemic has waned and hospital Trusts have placed greater emphasis on control of the infection. A 61% reduction occurred in the number of cases reported to the HPA between April 2007 and April 2010, and current reports show an overall population rate of infection of 3.9 cases per 10,000 population-years, with 2.6 reports per 10,000 bed-days [15,16].

Clinical Disease

Asymptomatic Carriage

Asymptomatic colonisation occurs in an estimated 1-3% of healthy adults, although use of antibiotics and hospitalisation has been demonstrated to increase this rate substantially [17]. Contamination of the environment with spores is believed to contribute to the nosocomial acquisition of colonisation, particularly in hospital and long-term residential settings [1,18].

Risk Factors for Infection

The development of symptomatic infection most often occurs after exposure of colonised individuals to antimicrobial therapy. Antibiotics disrupt the balance of bowel flora, and allow *C. difficile* to flourish in susceptible individuals. Clindamycin was the first antibiotic to be positively associated with the development of CDAD, but many studies have since implicated other antibiotics including penicillins, cephalosporins, fluoroquinolones, and antibiotics used for surgical prophylaxis [7,19]. Additional risk factors for development of active infection include advancing age; treatment with chemotherapy for malignancy [20]; immunosuppressant treatment following solid organ transplantation [21, 22]; and low antibody levels against toxin A (the levels of which have also been found to decline with age) [23,24]. The majority of cases of CDAD have had recent admission to hospital facilities [25], although there are documented reservoirs of infection within the community and the diagnosis should also be considered in those with no previous contact with medical services [26,27].

It is uncertain whether exposure to acid suppressants such as proton pump inhibitors (PPI) and H_2 receptor antagonists (H_2RA) may increase the risk of developing CDAD. The spores of *C. difficile* are relatively acid-stable, and this contributes to their pathogenicity, since they are less resistant to the acid environment within the stomach [28]. Theoretically, therefore, adjusting the pH should not have a significant effect upon colonisation and infection with the organism, and this is borne out in several studies which have not found any association between acid-suppressant medications and the development of CDAD[2,29]. However, other studies, including a systematic review of the literature, have reported an association between the infection and these medications [30,31], although it is not clear whether this association is causal.

Other reported associations with CDAD include; inflammatory bowel disease [32]; women in the peripartum stage [33]; advanced HIV disease [34] and cystic fibrosis. A high rate of asymptomatic colonisation rate has been described in this patient group, although the risk of developing clinical disease and severe colitis is elevated in those who have undergone lung transplantation [35].

Clinical Infection

The cardinal symptom of infection with *C. difficile* is profuse, watery diarrhoea. A spectrum of disease occurs, ranging from a rapidly resolving mild diarrhoeal illness, through to colitis and systemic upset, and at the most severe end, a fulminant pseudomembranous colitis with colonic dilatation (toxic megacolon) and a high mortality rate [5]. Affected individuals are predominantly elderly patients, often with other medical or surgical comorbidities.

Extracolonic Disease

C. difficile almost always affects the large bowel, but cases of small bowel enteritis following abdominal surgery, including total colectomy, have been reported. A review of the cases established that several patients had underlying inflammatory bowel disease, and that the presentation of disease was often severe, with a high mortality rate [36].

Documented extraintestinal manifestations of *C. difficile* infection include bone and joint infections (including prosthetic joint infection), skin and soft tissue infection, pancreatic and splenic abscesses and pleural empyema. Cases of reactive arthritis following CDAD are also well described [37-41].

The development of *C. difficile* bacteraemia is relatively uncommon, but has been reported in cases of severe gastrointestinal disease, and is postulated to arise from translocation of the organism across a pathologically disrupted bowel wall. There have been reports of bacteraemia in patients not previously known to have the infection, and these cases presumably represent translocation in colonised individuals. There has been one report to date of bacteraemia with the ribotype 027 strain [42,43].

Markers for Severity of Infection

A number of studies have aimed to elucidate markers of severe infection, for the purposes of stratifying patients during treatment. Table 1 lists several such markers, although there is no universally accepted case definition. A diagnosis of severe CDAD is made on an individual basis, after taking clinical, haematological/biochemical and radiological severity indicators into consideration.

Table 1. Markers of severity in *Clostridium difficile*- associated diarrhea

Total leukocyte count greater than 15,000 cells/mL [10]
Serum creatinine greater than 1.5 times the premorbid level [10]
Endoscopic appearances of Pseudomembranous colitis [44]
Admission to Intensive Care Unit [44]
Low serum albumin level [44,45]
Age over 60 years [44]
Systemic fever [44]

DIAGNOSIS

Laboratory Methods of Diagnosis

Diagnosis of *C. difficile* associated diarrhoea is made by the association of diarrhoea, and either a positive stool culture or positive test for toxins, or endoscopic or histopathological findings consistent with pseudomembranous colitis. Caution needs to be exercised when stool culture is positive but toxin testing is negative, as this could represent the presence of a non-pathogenic strain, and therefore colonisation rather than infection [46].

The "gold standard" for laboratory diagnosis is the culture and confirmation of a toxin-producing strain of the organism, although this is time-consuming and not feasible for every laboratory. Detection of toxin by enzyme immunoassay (EIA) techniques, are quicker to perform, but have reduced sensitivity (63-94%) and specificity (75-100%) in comparison with cytotoxigenic culture [10]. Despite this, they still comprise the main method of diagnosis in the majority of laboratories. EIA assays detect either toxin A alone, or both toxins A and B, since some strains which are A-/B+ do occur [47]. Some kits also contain a test for the glutamate dehyrogenase enzyme as an additional indicator of the presence of *C. dificile*.

The presence of toxins in the faeces is known to persist for some time following the resolution of clinical symptoms, and toxin identification is not generally recommended as a test of cure [10,48]. Likewise the development of a possible relapse of infection within the first few weeks of the initial episode is probably best diagnosed clinically by the recurrence of diarrhoea and other systemic symptoms, rather than upon retesting of stool for toxins.

Real-time PCR methods for the detection of *C. difficile* toxin genes are available in some clinical settings. PCR techniques include detection of tcd A and tcd B genes, as well as the deletional mutation in the tcdC gene (Δ117) which characterises the 027 strain [49,50,51]. The studies demonstrate good sensitivity and specificity values, possibly better than with enzyme immuno-assays, and one author suggests that their high negative predictive value may enable them to be used as a first negative screening step in the diagnostic process [49].

Strain differentiation by PCR ribotyping is available in some centres. Although typing does not alter the management of individual patients, or influence the implementation of infection control procedures, it does allow epidemiological study of the infection [47]. During the mid-2000s, strain typing allowed the epidemic 027 type strain to be followed as it spread across the USA and into Europe. Studies from the UK demonstrated that during the epidemic, type 027 replaced previous dominant strains such as 001. Since the waning of the epidemic, it has again been possible to demonstrate a shift in ribotype prevalence, with other strains again now contributing a greater proportion to the overall disease burden [52].

Radiological Diagnosis

Radiological techniques for diagnosis of CDAD complement clinical and laboratory findings in many cases, but have in some instances been the trigger for consideration of the disease in sick patients. Xrays of the abdomen may be entirely normal in mild infections, or demonstrate a variety of abnormalities depending upon the severity of the disease. A spectrum of small and large bowel dilatation (Figure 1), ranging up to toxic megacolon and perforation with pneumoperitoneum can occur. Haustral fold thickening is described as “thumb-printing”, and mucosal thickening has also been described, which is presumed to represent the presence of pseudomembranes on the surface of the mucosa [53]. Colonic thickening and ascites may also be seen on ultrasound examination [54].

Findings on computed tomography (CT) include thickening of the colonic walls, colonic dilatation, pericolonic oedema with stranding, and ascites in severe cases (Figure 2) [55,56]. An additional reported finding is that of the “accordion sign”, which occurs when oral contrast material is trapped between oedematous ridges of colonic mucosa and pseudomembrane plaques [57]. These radiological findings may also be seen in other conditions such as inflammatory bowel disease, bowel ischaemia and diverticulitis.

Clinical Management of Infection

The management of CDAD focuses on a number of different areas including; initial measures to prevent the emergence of infection; treatment of the individual affected patient; attempts at prevention of the relapse of infection; and limitation of the spread of infections to others.

Asymptomatic, colonised individuals do not need any specific treatment. However, the development of diarrhoea in a patient previously known to be colonised with the organism should raise a high index of suspicion for CDAD.

As a first measure, affected and clinically symptomatic individuals should have the causative antibiotic stopped if possible. If concomitant infections are present, the patient should be treated with antimicrobials which have a lesser propensity to cause CDAD.

Recommended antibiotic treatment against *C. difficile* varies depending upon whether a patient has severe or non-severe disease. As discussed above, the distinction between the two categories is usually made on a case by case basis, taking a number of different parameters into consideration (Table 1).

Antibiotic Treatment

For many years, oral metronidazole has been the first-line treatment for non-severe CDAD, largely due to concerns in the 1990s about the potential spread of vanomycin-resistant enterococci (VRE) [58]. More recently, it was reported that colonisation with VRE was unaffected by the use of vancomycin [59], whereas a further study demonstrated similar VRE colonisation rates with both vancomycin and metronidazole [60].

A Cochrane review of the literature in 2008 found that metronidazole, along with a number of other antibiotics including fusidic acid and nitazoxanide, was equivalent in efficacy to oral vancomycin. Teicoplanin was shown to be more effective than vancoymcin, but is unavailable in the USA, and this, together with its significantly higher cost, limits its overall use [61]. However, since the literature search for the Cochrane review was performed, a randomised controlled trial reported that orally-administered vancomycin was significantly better than metronidazole at effecting a clinical cure in patients with severe disease. Clinical cure occurred in 76% of those taking metronidazole, versus 97% of the patients taking vancomycin. Severe disease was defined according to a scoring system devised for the trial, although several widely accepted severity markers were included [44]. In this trial, both vancomycin and metronidazole were equally effective for the patients with mild CDAD.

Minimum inhibitory concentrations (MIC) to both metronidazole and vancomycin have generally remained low [10, 62, 63], meaning that in vitro, the organism is susceptible to both antibiotics. The epidemic NAP1/B1/027 strain was found to be resistant to quinolones, but retained susceptibility to both metronidazole and vancomycin [2,47]. However, studies demonstrating better clinical response to vancomycin have attributed this to the fact that there is virtually no systemic absorption of orally administered vancomycin; almost

100% remains within the bowel. In contrast, there is significant absorption of orally administered metronidazole in the first part of the bowel, thereby leaving a smaller fraction to act upon the colon.

Following the recent randomised controlled trial, the recommendation for the first-line treatment of severe disease is oral vancomycin at a dose of 125mg four times daily. Although higher doses have been used, some studies have demonstrated clinical equivalence [64], presumably due to the fact that even at the lower dose, very high levels are found within the colon. However, since ileus may reduce the amount of orally-administered antibiotic which reaches the colon, intravenous metronidazole may be an alternative strategy. The use of intracolonic vancomycin has been reported, although it is unclear how much of this drug, administered via a bowel management tube, actually reaches the right side of the colon. This approach has not been studied in any formal trial situation, although there are various case reports and series which seem to indicate that it may be of use in some patients [65,66]. In most reports, a dose of 2g per day in divided doses has been administered. Intracolonic vancomyin is recommended as possible adjunctive therapy in severe episodes of CDAD complicated by ileus, in the 2008 Infectious Diseases Society of America (IDSA) guidelines on *C. difficile* [10].

There is no evidence at present to support the administration of combination antibiotic therapy [61]. There is recently reported data on fidaxomicin, a macrocyclic antibiotic which is bacteriocidal against *C. difficile*. Patients with fulminant disease were excluded from the double-blind, randomised, parallel-group trial, in which fidaxomicin was compared with vancomycin. Fidaxomicin was shown to have similar clinical efficacy to vancomycin, but resulted in fewer relapses [67].

Surgical Treatment

In a minority of patients, antibiotic treatment is insufficient to prevent progression to fulminant disease, and surgical intervention is required. Indications for surgery are the development of clinical signs consistent with peritonitis and perforation, or radiological evidence of worsening colonic toxic dilatation [68,55]. The treatment of choice is a total colectomy and ileostomy formation, since the degree of pancolitis present usually precludes the ability to perform a colonic anastomosis [57]. Colectomy has a more favourable outcome than more limited surgical procedures [55].

Relapsed Disease

Recurrent episodes of CDAD, (defined as the re-emergence of symptoms following complete abatement of diarrhoea and other symptoms after treatment), may be due either to reinfection with a new strain, or due to relapse of infection [69]. Spores within colonic diverticulae, for example, may withstand the peristaltic action of the bowel, and subsequently revegetate, causing relapse of disease. There are a number of factors which have been identified as contributing to an increased risk of relapse following the initial CDAD episode. These include; advancing age; underlying medical co-morbidities; use of antibiotics following the first episode [45]; and a poor immune response against Toxin A [23].

Alternative causes for recurrent diarrhoea should be considered, such as inflammatory bowel disease, other gastrointestinal infections and malignancy. Lower gastrointestinal endoscopy may be warranted to exclude these other diagnoses.

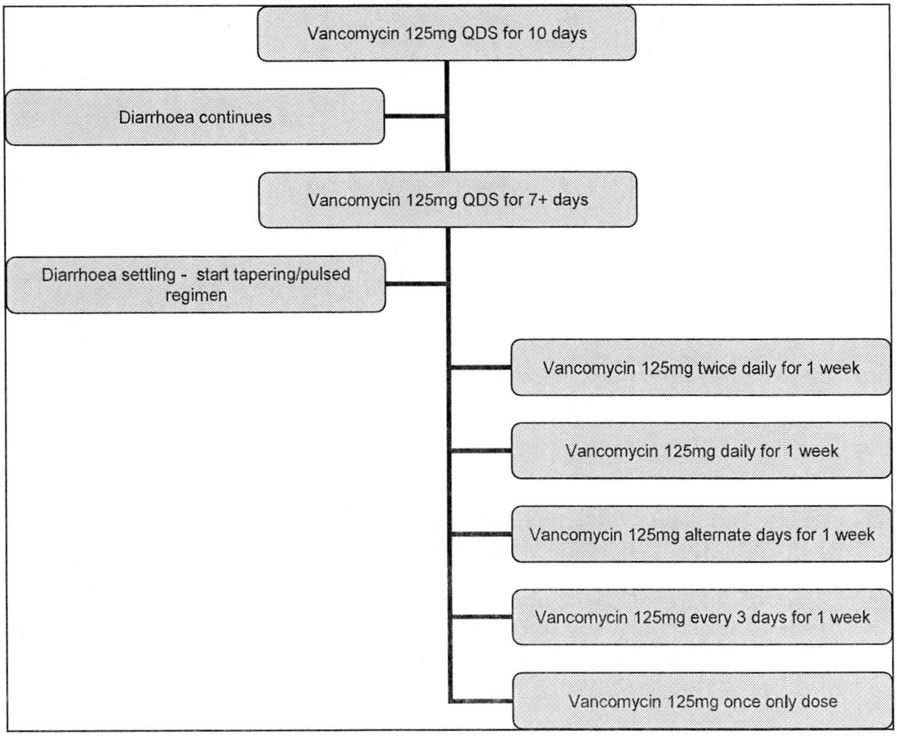

Figure 3. Suggested tapering/pulsed Vancomycin regimen.

Treatment of recurrent episodes generally involves re-treatment with either metronidazole or vancomycin. There are varying different strategies; some clinicians use the treatment which resulted in resolution of symptoms on the previous occasion; others use the alternative drug. Tapering and pulsed, or intermittent vancomycin therapy courses have each been used. Tapering therapy involves gradually decreasing the dose of vancomycin, usually over several weeks, irrespective of whether diarrhoea resolves in the interim or not. Pulsed therapy involves administering vancomycin doses on alternate days, or every few days. The rationale behind this approach is that it allows residual spores within the bowel to germinate inbetween doses; these new vegetative organisms are then targeted when the next dose is administered. In addition, this strategy may allow the normal bowel ecoflora to repopulate on the days when antibiotics are omitted. Several studies have demonstrated a reduced rate of subsequent relapse [70] when such strategies are implemented. Tapering and pulsed therapies may be combined in a single regimen. Figure 3 illustrates one possible regimen as used in our hospital. These strategies may also be used in a first episode of CDAD, when diarrhoea is slow to fully resolve after a standard 10 or 14 day course of treatment.

Faecal Biotherapy

Successful resolution of CDAD depends not only on eradication of the organism, but on reconstitution of the normal bowel flora population in order to suppress the recurrent growth of *C. difficile* [71,72]. Patients who have had recurrent episodes of CDAD, or who have failed to resolve their initial infection, may have difficulty in achieving the normal bowel repopulation. It is unclear which of the normal bowel flora may be protective against *C. difficile* [48]. Several case series have reported the use of faecal biotherapy, which involves the administration of faeces from a healthy donor to the patient, either through an enema [73] or colonoscope [71], or via a nasogastric [48,72] or nasojejunal tube [74]. The earliest report of this technique for pseudo-membranous colitis was actually in 1958, even before the formal recognition of *C. difficile* as the offending pathogen [73].

In faecal biotherapy, the donor is usually a relative of the patient, for two reasons; firstly, because this may be more acceptable to the patient, and secondly, because there is an assumption that people living in the same household, eating similar food, have a similar bowel ecoflora pattern [71]. The donor should be screened for blood borne viruses, including HIV, although the risk of transmission of these viruses through faeces is almost negligible. Donor faeces is also cultured for pathogenic bacteria, examined for parasites, ova and cysts, and tested for *C. difficile* toxin, prior to administration. The usual preparation involves a fresh stool specimen which is blended with saline and filtered to remove large particles. A small volume of the remaining liquid (approximately 20ml) is administered [72]. The treatment time is short, and the procedure has been performed in outpatient settings with good results [72], as well as in patients who are dependent upon inpatient care [48].

Although the evidence for faecal biotherapy has so far come solely from relatively small case series, a randomised trial based in the Netherlands is currently underway, comparing the technique with conventional antibiotic therapy in cases of relapsed infection.

Nutrition

Many patients with CDAD are elderly, debilitated individuals with underlying medical co-morbidities. The need for nutritional support and to repopulate the bowel with normal resident flora is crucial. Symptoms such as diarrhoea, nausea and abdominal discomfort, together with hospitalisation, and being removed from family and familiar food, can all contribute to a reduction in the desire to eat. In our CDAD cohort/isolation unit, the involvement of the dietician as part of the multi-disciplinary team is of particular importance. We employ the use of both nutritional dietary supplements, as well as nasogastric feeding for patients who find it difficult to maintain an adequate intake. The published literature on the use of adjuvant enteral feeding for patients with CDAD is extremely scant, since most reviews focus on the antimicrobial aspect of care. However, there has been much attention paid to the overall nutritional status of all patients in hospital, particularly the elderly, and addressing malnutrition should be an integral part of in-patient care [75].

Alternative Treatments

Probiotic Treatment

Probiotics are live microorganisms which are administered therapeutically, either as pure cultures or contained within food such as yogurt. They have been marketed widely as being of benefit to patients with a variety of health needs, with the aim of "restoring the natural microbial balance" [76]. A few prospective randomised trials have examined the use of probiotic treatments for CDAD, either as sole agents, or as adjunctive therapy with antibiotics [77-80]. However, a Cochrane review of the published literature concluded that there was currently insufficient current evidence to recommend their use in the management of *C. difficile* [81]. In addition, caution should be taken when administering these compounds to severely debilitated patients, as fungaemia with these organisms has been reported in the context of treatment for CDAD [82].

Intravenous Immunoglobulin Therapy

Intravenous immunoglobulin (IVIG) therapy has been the focus of several papers reporting upon the treatment of CDAD. Declining levels of antibody against toxin A is a risk factor for the development of clinical disease, and specific antibody levels have also been shown to decline with advancing age [23,24].The passive infusion of anti-toxin A IgG antibodies leads to the neutralisation of toxin A, and in animal models, has resulted in control of infection. In humans, there have been a number of reports, all non-randomised and mainly of a retrospective nature, which have demonstrated clinical benefit of IVIG [83-87]. However, due to the lack of randomised controlled trials, there is insufficient evidence at present to recommend IVIG therapy in the management of CDAD [83,85].

Anion-Binding Resins

Tolevamer, a non-antimicrobial polystyrene resin, postulated to neutralise toxins A and B, showed promising results in an initial phase II trial [88]. However, the compound was later compared with vancomycin in phase III trials and failed to meet the non-inferiority endpoint [89,90]. Subsequent data using an in vitro human gut model failed to demonstrate neutralisation of toxins [91]. There is therefore currently no evidence for the role of tolevamer in the treatment of CDAD.

Monoclonal Antibodies

Monoclonal antibodies targeting toxins A and B are under current investigation for clinical use, following studies which correlated higher antibody levels against toxin A with decreased recurrence of CDAD [23]. Following successful animal modelling, a phase II double-blind, placebo controlled trial was published in 2010. The authors reported a reduction in the development of recurrence of CDAD amongst patients who received the antibodies in conjunction with standard antimicrobial therapy, in comparison with patients who received standard therapy alone [92]. Further research is needed before this treatment can be formally recommended as an adjunct to current practice.

Vaccination

Several different vaccines have been tested in both animals and humans [93,94] but none has yet been licensed for general administration. At the present time a phase II randomised, placebo-controlled trial of a toxoid vaccine is in progress, with a primary endpoint measure of recurrence of infection. The trial is expected to end in 2011.

Prognosis and Mortality

The prognosis in patients with uncomplicated CDAD is good, with the majority of cases settling with a first line treatment of either oral metronidazole or vancomycin. In our experience, the patients who improve most quickly are those who have few or no co-morbidities, with good underlying performance status and adequate nutrition.

Severe disease is associated with significant morbidity including dehydration and electrolyte dysfunction as a result of volume depletion from profuse diarrhoea, organ failure and sepsis [2,3]. Hypoalbuminaemia and resultant malnutrition is common, and may limit the chances of recovery even when the infectious episode is controlled. Toxic megacolon and bowel perforation may result from severe colitis and are associated with significant morbidity and mortality, especially in the elderly, for whom the risks of surgical intervention may be deemed unacceptable by the patient and medical staff.

Mortality from CDAD varies according to the severity of the infection, and the underlying health of those affected. Published literature produced prior to the emergence of the endemic strain demonstrated variable mortality rates for CDAD. Observed crude mortality rates in observational and case control studies were reported from under 1% to above 60% [65, 95-98]. More recent publications have reported an appreciable rise in mortality associated with the 027-endemic strain [2,13] although studies are conflicting, with some reports suggesting that attributable mortality remains generally low [99]. A review of 27 studies reporting mortality outcomes from before and during the endemic strain, found that there was an overall mortality rate of 5.99% associated with the infection, and that higher rates were observed in the very elderly, and in studies published since the year 2000 [100]. Other than advancing age, factors associated with this rise in mortality include the presence of the more virulent strain, and possible increased reporting of associated deaths by institutions.

In the UK, the infection can be reported on the death certificate as either the overall cause of death, or as a contributory factor, and this difference may need to be taken into account if mortality rates are extrapolated directly from certificate data. Barriers to accurate reporting may include possible stigmatisation of individual institutions, and increased workload from preparation of root cause analyses and Coroner's inquests [101].

INFECTION CONTROL POLICIES

Infection control measures include those which underpin efforts to prevent the emergence of infection within a particular environment, such as a hospital, as well as the immediate response to the detection of a clinical case. Both the IDSA, and the HPA in the UK have published evidence-based guidelines which address infection control issues in healthcare settings, and both are referred to here [10,47].

The chance of becoming colonised with *C. difficile* increases with prolonged duration of hospital stay, through contact with infected patients or a contaminated environment. There is the potential for healthcare staff to contribute to the spread of infection, by carrying spores on their hands from an infected case to other patients or to different areas of the ward or institution [10]. Hand hygiene is therefore an area upon which many institutions have focussed during their efforts to control CDAD. The spores of *C. difficile* are largely unaffected by alcohol-based preparations [102], and good handwashing techniques using soap and water are the recommended measure for all healthcare staff [10,47] to prevent the spread of infection. Affected individuals should be isolated if possible, or cohorted with other cases if there is a large outbreak of infection. Standard contact precautions should be implemented, including disposable gown and glove use, and careful disposal of faecal material [10,47].

Regular environmental cleaning is essential, since the spores of *C. difficile* can remain on many surfaces within a healthcare setting; their hardy resistance to dessication means they can survive for many months in this form, and in addition they are able to withstand the action of many commonly used biocides [103]. Chlorine has been the only reliable sporocidal agent [104], and products with a minimum of 1,000 ppm available chlorine are the cleaning agents of choice for ward areas and patient bathrooms [10,47].

The vast majority of patients who develop clinical disease with *C. difficile* have had antibiotic prescriptions in the recent past, either as an inpatient or in the community [25]. The implementation of an antibiotic stewardship programme is one of a number of measures which have been taken by hospitals in order to minimise the prescribing of antibiotics which are considered "high risk" for the development of CDAD [7,105,106]. Such programmes usually involve the creation of a list of restricted antimicrobials for which authorisation is required from a microbiologist or infectious diseases physician, before they can be administered to patients. Examples include restriction on the use of third-generation cephalosporins for certain community acquired infections, and limitation of the use of quinolones, which were associated with the emergence of the epidemic 027 strain. In addition to restriction of the use of individual antimicrobials, guidelines have been issued which aim to prompt clinicians to improve their overall antibiotic prescribing; including regular reviewing of charts; documentation of duration of therapy; and early switching to narrow-spectrum agents where possible [25,47].

It is often difficult to assess the contribution of each factor to the control of infection, as most centres have used a "care bundle" approach to the control of *C. difficile*, incorporating several interventions. This is the approach recommended by the UK Department of Health and implementation of these control measures has led to a spectacular decline in the number of cases in several centres [105].

Conclusion

Diarrhoea associated with *C. difficile* is a common nosocomial infection, predominantly associated with the prior use of antibiotics for unrelated infections. For many, the infection leads to a debilitating and miserable illness, and in complicated cases, increases the mortality of the most elderly and vulnerable patients. Management focuses on the use of antimicrobial therapy, and is best managed by an enthusiastic multi-disciplinary team, with emphasis on nutrition and rehabilitation. Continued research into new therapies such as vaccination and other immunomodulatory therapies is needed. Consolidation of infection control procedures, continued support from management steering groups, and heightened awareness amongst medical and nursing staff is needed if further rises in incidence are to be averted.

References

[1] McFarland LV, Mulligan ME, Kwok RY, Stamm WE. Nosocomial acquisition of Clostridium difficile infection *N Engl J Med.* 1989;320(4):204.

[2] Loo VG, Poirier L, Oughton M, Libman MD, Michaud S, Bourgault A-M et al. A Predominantly Clonal Multi-Institutional Outbreak of *Clostridium-difficile* Associated Diarrhea with High Morbidity and Mortality. *NEJM* 2005; 353: 2442-2449.

[3] McDonald LC, Killgore GE, Thompson A, Owens RC, Kazakova SV et al. An Epidemic Toxin Gene-Variant Strain of *Clostridium difficile.* New England Journal of Medicine 2005; 353 (23): 2433-2441.

[4] Hall, IC, O'Toole, E. Intestinal flora in newborn infants with a description of a new pathogenic anaerobe Bacillus difficilis. *Am J Dis Child* 1935; 49:390

[5] Hurley BW, Nguyen CC. The Spectrum of Pseudomembranous Enterocolitis and Antibiotic-Associated Diarrhea. *Archives of Internal Medicine* 2002; 162: 2177-2184

[6] George RH, Symonds JM, Dimock F, Brown JD, Arabi Y, Shinagawa N, Keighley MR, Alexander-Williams J, Burdon DW. Identification of Clostridium difficile as a cause of pseudomembranous colitis.*Br Med J.* 1978;1(6114):695.

[7] Spencer RC. The role of antimicrobial agents in the aetiology of *Clostridium difficile-* associated diarrhoea. *Journal of Antimicrobial Chemotherapy* 1998; 41: Suppl C, 21-27.

[8] He, D., S. Sougioultzis, S. Hagen, J. Liu, S. Keates, A. C. Keates, C. Pothoulakis, and J. T. Lamont. 2002. *Clostridium difficile* toxin A triggers human colonocyte IL-8 release via mitochondrial oxygen radical generation. *Gastroenterology* 122:1048–1057.

[9] Voth DE, Ballard JD. *Clostridium difficile* Toxins: Mechanism of Action and Role in Disease. *Clinical Microbiology Reviews* 2005; 18(2): 247-263.

[10] Cohen SH, Gerding DN, Johnson S, Kelly CP, Loo VG, McDonald LC, Pepin J, Wilcox MH. Clinical practice guidelines for Clostridium difficile infection in adults: 2010 update by the society for healthcare epidemiology of America (SHEA) and the infectious diseases society of *America. Infect Control Hosp Epidemiol.* 2010;31(5):431.

[11] Warny M, Pepin J, Fang A et al. Toxin production by an emerging strain of *Clostridium difficile* associated with outbreaks of severe disease in North America and Europe. *Lancet* 2005; 366: 1079-1084.

[12] Schwan C, Stecher B, Tzivelekedis T, van Ham M, Rohde M, Hardt W-D et al. *Clostridium difficile* toxin CDT induces formation of microtubule-based protrusions and increases adherence of bacteria. *PLoS Pathog* 5(10): e1000626. doi:10.1371/journal.ppat.1000626

[13] Redelings MD, Sorvillo F, Mascola L. Increase in *Clostridium difficile*-related mortality rates, United States, 1999–2004. *Emerg Infect Dis* 2007;13:1417–1419.

[14] Health Protection Agency. *Surveillance of Healthcare associated infections report 2008*. London: Health Protection Agency. 2008

[15] Health Protection Agency 2010. MRSA bacteraemia and *Clostridium difficile* mandatory reports 2009/10. Available from *http://www.hpa.org.uk/Topics/InfectiousDiseases/InfectionsAZ/ClostridiumDifficile*

[16] Health Protection Agency 2011. *Quarterly Analyses: Mandatory MRSA and MSSA bacteraemia and CDI infection in England* (up to January-March 2011). London: Health Protection Agency, June 2011.

[17] McMaster-Baxter NL, Musher DM. Clostridium difficile: recent epidemiologic findings and advances in therapy. *Pharmacotherapy.* 2007;27:1029–39.

[18] Riggs MM, Sethi AK, Zabarsky TF, Eckstein EC, Jump RL, Donskey CJ. Asymptomatic carriers are a potential source for transmission of epidemic and nonepidemic *Clostridium difficile* strains among long-term care facility residents. *Clinical Infectious Diseases* 2007;45(8):992.

[19] Muto CA, Pokrywka M, Shutt K, Mendelsohn AB, Nouri K, Posey K et al. A large outbreak of *Clostridium difficile* – associated disease with an unexpected proportion of deaths and colectomies at a teaching hospital following increased fluroquinolone use. Infection Control and Hospital *Epidemiology* 2005; 26: 273-280.

[20] Anand A, Glatt AE.Clostridium difficile infection associated with antineoplastic chemotherapy: a review. *Clin Infect Dis.* 1993 Jul;17(1):109-13.

[21] Stelzmueller I, Goegele H, Biebl M, Wiesmayr S, Berger N, Tabarelli W, Ruttmann E, Albright J, Margreiter R, Fille M, Bonatti H. *Clostridium difficile* colitis in solid organ transplantation--a single-center experience. *Dig Dis Sci.* 2007;52(11):3231-6.

[22] Schaier M, Wendt C, Zeier M, Ritz E. *Clostridium difficile* diarrhoea in the immunosuppressed patient – update on prevention and management. *Nephrol Dial Transplant* 2004; 19: 2432-2436.

[23] Kyne L, Warny M, Qamar A, Kelly CP. Association between antibody response to toxin A and protection against recurrent *Clostridium difficile* diarrhoea. *Lancet* 2001; 357: 189-193.

[24] Bacon AE 3rd, Fekety R. Immunoglobulin G directed against toxins A and B of Clostridium difficile in the general population and patients with antibiotic-associated diarrhea. *Diagn Microbiol Infect Dis.* 1994 Apr;18(4):205-9.

[25] Chang HT, Krezolek D, Johnson S, Parada JP, Evans CT, Gerding DN. Onset of symptoms and time to diagnosis of *Clostridium difficile*-associated disease following discharge from an acute care hospital. *Infect Control Hosp Epidemiol.* 2007; 28(8): 926-31.

[26] Riley TV, Cooper M, Bell B, Golledge CL. Community- acquired Clostridium difficile–associated diarrhea. *Clin Infect Dis* 1995;20 (suppl 2):S63-S265.

[27] al Saif N, Brazier JS. The distribution of Clostridium difficile in the environment of South Wales. *J Med Microbiol* 1996;45:133-137.

[28] Gurian L, Ward TT, Katon RM. Possible foodborne transmission in a case of pseudomembranous colitis due to *Clostridium difficile*: influence of gastrointestinal secretions on *Clostridium difficile* infection. *Gastroenterology*. 1982 Aug;83(2):465-9.

[29] Naggie S, Miller BA, Zuzak KB, Pence BW, Mayo AJ, Nicholson BP et al. A case-control study of community- associated *Clostridium difficile* infection: no role for proton pump inhibitors. *Americal Journal of Medicine* 2011; 124(3): 276.e1-7.

[30] Linsky A, Gupta K, Lawler EV, Fonda JR, Hermos JA. Proton pump inhibitors and risk for recurrent *Clostridium difficile* infection. *Archives of Internal Medicine* 2010; 170(9): 772-8.

[31] Leonard J, Marshall JK, Moayyedi B. Systematic review of the risk of enteric infection in patients taking acid suppression. *American Journal of Gastroenterology* 2007; 102: 2047-2056.

[32] Rodemann JF, Dubberke ER, Reske KA, Seo da H, Stone CD. Incidence of *Clostridium difficile* infection in inflammatory bowel disease. *Clin Gastroenterol Hepatol.* 2007;5(3):339-44.

[33] Rouphael NG, O'Donnell JA, Bhatnagar J, Lewis F, Polgreen PM, Beekmann S et al. *Clostridium difficile*- associated diarrhoea: an emerging threat to pregnant women. *American Journal of Obstetrics and Gynecology* 2008; 198: 635.e1-635.e6.

[34] Sanchez TH, Brooks JT, Sullivan PS, Juhasz M, Mintz E, Dworkin MS et al. Bacterial Diarrhea in in Persons with HIV Infection, United States, 1992-2002. *Clinical Infectious Diseases* 2005; 41: 1621-1627.

[35] Theunissen C, Knoop C, Nonhoff C, Byl B, Claus M, Leisnard C et al. *Clostridium difficile* colitis in cystic fibrosis patients with and without lung transplantation. *Transplant Infectious Diseases* 2008; 10(4): 240-244.

[36] Freiler JF, Durning SJ, Ender PT. *Clostridium difficile* Small Bowel Enteritis occurring after Total Colectomy. *Clinical Infectious Diseases* 2001; 33: 1429-141431.

[37] Pron B, Merckx J, Touzet P, Ferroni A, Poyart C, Berche P et al. Chronic septic arthritis and osteomyelitis in a prosthetic knee joint due to *Clostridium difficile. European Journal of Clinical Microbiology and Infectious Diseases* 14(7): 599-601

[38] Sofianou DC. Pancreatic abscess caused by *Clostridium difficile. European Journal of Clinical Microbiology and Infectious Diseases* 7(4): 528-529.

[39] Kumar N, Flanagan P, Wise C, Lord R. Splenic abscess caused by *Clostridium difficile. European Journal of Clinical Microbiology and Infectious Diseases* 16(12): 938-939.

[40] Extraintestinal *Clostridium difficile*: 10-years' experience at a tertiary-care hospital. *Mayo Clin Proc* 1998; 73(10): 943-7.

[41] Jacobs A, Barnard K, Fishel R, Gradon JD. Extracolonic manifestations of *Clostridium difficile* infections. *Medicine* 2001; 80: 88-101.

[42] Hemminger J, Balada-Llasat J, Raczkowski M, Buckosh M, Pancholi P. Two case reports of Clostridium difficile bacteremia, one with the epidemic NAP-1 strain. *Infection* (E-pub ahead of print) DOI 10.1007/s15010-011-0115-7.

[43] Libby DB, Bearman G. Bacteremia due to Clostridium difficile—review of the literature. *Int J Infect Dis*. 2009;13:e305–9.

[44] Zar FA, Bakkanagari SR, Moorthi KMLST, Davis MB. A comparison of vancomycin and metronidazole for the treatment of *Clostridium difficile*-associated diarrhea, stratified by disease severity. *Clinical Infectious Diseases* 2007; 45: 302-307.

[45] Nair S, Yadav D, Corpuz M, Pitchumini CS. *Clostridium difficile* colitis: Factors influencing treatment failure and relapse – A prospective evaluation. *Americal Journal of Gastroenterology* 1998; 93(10): 1873-1876.

[46] Gerding DN, Johnson S, Peterson LR, Mulligan ME, Silva J Jr. *Clostridium difficile*-associated diarrhea and colitis. *Infect Control Hosp Epidemiol.* 1995 Aug;16(8):459-77.

[47] Health Protection Agency. *Clostridium difficile* infection: How to deal with the problem 2008. Health Protection Agency and the Department of Health. Available from: *http://www.hpa.org.uk/Topics/InfectiousDiseases/InfectionsAZ/ClostridiumDifficile/Guidelines/*

[48] MacConnachie AA, Fox R, Kennedy DR, Seaton RA. Faecal transplant for *Clostridium difficile*- associated diarrhoea: a UK case series. *Q J Med* 2009; 102: 781-784.

[49] Knetsch CW, Bakker D, de Boer RF, Sanders I, Hofs S, Kooistra-Smid AMD et al. Comparison of real-time PCR techniques to cytotoxigenic culture methods for diagnosing *Clostridium difficile* infection. *Journal of Clinical Microbiology* 2011; 49(1): 227-231.

[50] Sloan LM, Duresko BJ, Gustafson DR, Rosenblatt JE. Comparison of real-time PCR for the detection of the tcdC gene with four toxin immunoassays and culture in diagnosis of *Clostridium difficile* infection. *Journal of Clinical Microbiology* 2008; 46(6): 1996-2001.

[51] Peterson LR, Manson RU, Paule SM, Hacek DM, Robicsek A, Thomson RB. Detection of toxigenic *Clostridium difficile* in stool samples by real-time polymerase chain reaction for the diagnosis of *C. difficile*-associated diarrhea. *Clinical Infectious Diseases* 2007; 45: 1152-1160.

[52] Health Protection Agency 2011. *Clostridium difficile* Ribotyping Network (CDRN) for England and Northern Ireland. Annual Report 2009/10. Available at: *http://www.hpa.org.uk/ProductsServices/InfectiousDiseases/LaboratoriesAndReferenceFacilities/ClostridiumDifficileRibotypingNetworkService/*

[53] Kawamoto S, Horton KM, Fishman EK. Pseudomembranous colitis: spectrum of imaging findings with clinical and pathological correlation. *RadioGraphic* 1999; 19: 887-897.

[54] Sheridan JS, Hunt G, MacLennan AC. Ultrasound diagnosis of clinically undetected *Clostridium difficile* toxin colitis. *Clinical Radiology* 2006; 61: 902.

[55] Lipsett PA, Samantaray DK, Tam ML, Bartlett JG, Lillemoe KD. Pseudomembranous colitis: a surgical disease? *Surgery*. 1994 Sep;116(3):491-6.

[56] Valiquette L, Pepin J, Do X-V, Nault V, Beaulieu A-A, Bedard J, Schmutz G. Prediction of complicated *Clostridium difficile* infection by pleural effusion and increased wall thickness on computed tomography. *Clinical Infectious Diseases* 2009; 49: 554-560.

[57] Macari M, Balthazar EJ, Megibow AJ. The accordion sign at CT: A non-specific finding in patients with colonic edema. *Radiology* 1999; 211: 743-746

[58] Recommendations for preventing the spread of vancomycin resistance: Recommendations of the Hospital Infection Control Practices Advisory Committee (HICPAC). *MMWR Recomm Rep* 1995; 44(RR-12): 1-13.

[59] Miller M, Bernard L, Thompson M, Grima D, Pepin J. Lack of Increased Colonization with Vancomycin-Resistant Enterococci during Preferential use of Vancomycin for Treatment during an Outbreak of Healthcare Associated *Clostridium difficile* Infection. *Infection Control and Hospital Epidemiology* 2010; 31(7): 710-715.

[60] Al-Nassir WN, Sethi AK, Li Yuejin, Pultz MJ, Riggs MM, Donskey CJ. Both Oral Metronidazole and Oral Vancomycin Promote the Overgrowth of Vancomycin- Resistant Enterococci during Treatment of *Clostridium difficile*-associated Disease. *Antimicrobial Agents and Chemotherapy* 2008; 52(7): 2403-2406.

[61] Nelson R. Antibiotic treatment for *Clostridium difficile*-associated diarrhea in adults. *Cochrane Database of Systematic Reviews* 2007, Issue 3. Art. No.: CD004610. DOI: 10.1002/14651858.CD004610.pub3.

[62] Freeman J, Stott J, Baines SD, Fawley WN, Wilcox MH. Surveillance for resistance to metronidazole and vancoymcin in genotypically distinct

and UK epidemic *Clostridium difficile* isolates in a large teaching hospital. *Journal of Antimicrobial Chemotherapy* 2005; 56: 988-989.

[63] Drummond L, McCoubrey J, Smith DGE, Starr JM, Poxton IR. Changes in sensitivity patterns to selected antibiotics in *Clostridium difficile* in geriatric in-patients over an 18-month period. *Journal of Medical Microbiology* 2003; 52: 259-263.

[64] Fekety R, Silva J, Kauffman C, Buggy B, Deery HG. Treatment of antibiotic-associated Clostridium difficile colitis with oral vancomycin: comparison of two dosage regimens. *Am J Med.* 1989;86(1):15

[65] Olson MM, Shanholtzer CJ, Lee JT Jr, Gerding DN.Ten years of prospective Clostridium difficile-associated disease surveillance and treatment at the Minneapolis VA Medical Center, 1982-1991. *Infect Control Hosp Epidemiol.* 1994 Jun;15(6):371-81.

[66] Apisarnthanarak A, Razavi B, Mundy LM. Adjunctive intracolonic vancomycin for severe Clostridium difficile colitis: case series and review of the literature. *Clin Infect Dis.* 2002;35(6):690.

[67] Louie TJ, Miller MA, Mullane KM, Weiss K, Lentnek A, Golan Y et al. Fidaxomicin versus Vancomycin for *Clostridium difficile* infection. *New England Journal of Medicine* 2011; 364: 422-31.

[68] Koss K, Clark MA, Sanders DSA, Morton D, Keighley MRB, Goh J. The outcome of surgery in fulminant *Clostridium difficile* colitis. *Colorectal disease* 2006; 8(2): 149-154.

[69] Barbut F, Richard A, Hamadi K, Chomette V, Burghoffer B, Petit J. Epidemiology of Recurrences or Reinfections of *Clostridium difficile* – associated diarrhea. *Journal of Clinical Microbiology* 2000; 38(6): 2386-2388.

[70] McFarland LV, Elmer GW, Surawicz CM. Breaking the cycle; Treatment strategies for 163 cases of recurrent *Clostridium difficile* disease. *American Journal of Gastroenterology* 2002; 97(7): 1769-1775.

[71] Rohlke F, Surawicz CM, Stollman N. Fecal Flora Reconstitution for Recurrent *Clostridium difficile* Infection: Results and Methodology. J *Clin Gastroenterol* 2010; 44: 567-570

[72] Aas J, Gessert CE, Bakken JS. Recurrent *Clostridium difficile* colitis: Case series involving 18 patients treated with donor stool administered via a nasgogastric tube. *Clinical Infectious Diseases* 2003; 36: 580-585.

[73] Eiseman B, Silen W, Bascom GS, Kauvar AJ. Fecal enema as an adjunct in the treatment of pseudomembranous enterocolitis. *Surgery.* 1958 Nov;44(5):854-9.

[74] Grehan MJ, Borody TJ, Leis SM, Campbell J, Mitchell H, Wettstein A. Durable alteration of the colonic microbiota by the administration of donor fecal flora. *J Clin Gastroenterol.* 2010 Sep;44(8):551-61.

[75] Saunders J, Smith T. Malnutrition: causes and consequences. Clinical Medicine 2010; 10(6): 624-7.

[76] Elmer GW. Probiotics: "living drugs". *Am J Health Syst Pharm* 2001; 58(12): 1101-1109.

[77] Surawicz CM, McFarland LV, Greenberg RN, Rubin M, Fekety R, Mulligan ME et al. The search for a better treatment for recurrent *Clostridium difficile* disease: use of high-dose vancomycin combined with Saccharomyces boulardii. *Clinical Infectious Diseases* 2000; 31(4): 1012-7.

[78] McFarland LV, Surawicz CM, Greenberg RN, Fekety R, Elmer GW, Moyer KA et al. A randomised placebo-controlled trial of Saccharomyces boulardii in combination with standard antibiotics for *Clostridium difficile* disease. *JAMA* 1994; 271(24): 1913-1918.

[79] Wullt M, Hagslatt ML, Odenholt I. Lactobacillus plantarurn 299v for the treatment of recurrent *Clostridium difficile*- associated diarrhoea; a double-blind, placebo-controlled trial. *Scandinavian Journal of Infectious Diseases* 2003; 35(6-7): 365-7.

[80] Lawrence SJ, Korzenik JR, Mundy LM. Probiotics for recurrent *Clostridium difficile* disease. *J Med Microb* 2005; 54 (Pt 9): 905-6.

[81] Pillai A, Nelson RL. Probiotics for treatment of *Clostridium difficile* - associated colitis in adults. *Cochrane Database of Systematic Reviews* 2008, Issue 1. Art No. CD004611. DOI: 10.1002/14651858.CD004611.pub2.

[82] Cherifi S, Robberecht J, Miendje Y. Saccharomyces cerevisiae fungemia in an elderly patient with Clostridium difficile colitis. *Acta Clin Belg.* 2004;59(4):223-4.

[83] Hassoun A, Ibrahim F. Use of intravenous immunoglobulin for the treatment of severe Clostridium difficile colitis. *Am J Geriatr Pharmacother.* 2007 Mar;5(1):48-51.

[84] McPherson S, Rees CJ, Ellis R, Soo S, Panter SJ. Intravenous immunoglobulin for the treatment of severe, refractory, and recurrent Clostridium difficile diarrhea. *Dis Colon Rectum.* 2006 May;49(5):640-5.

[85] Beales ILP. Intravenous immunoglobulin for recurrent *Clostridium difficile* diarrhoea. *Gut* 2002; 51(3): 456.

[86] Juang P, Skledar SJ, Zgheib NK, Paterson DL, Vergis EN, Shannon WD, Ansani NT, Branch RA. Clinical outcomes of intravenous immune globulin in severe clostridium difficile-associated diarrhea. *Am J Infect Control.* 2007 Mar;35(2):131-7.

[87] Abougergi MS, Kwon JH. Intravenous Immunoglobulin for the Treatment of *Clostridium difficile* Infection: A Review. *Dig Dis Sci* 2011;56: 19-26.

[88] Louie TJ, Peppe J, Watt CK, Johnson D, Mohammed R, Dow G. Tolevamer, a novel nonantibiotic polymer, compared with Vancomycin in the treatment of mild to moderately severe *Clostridium difficile* – associated diarrhea. *Clinical Infectious Diseases* 2006; 43: 411-420.

[89] Louie T, Gerson M, Grimard D, et al. Results of a phase III trial comparing tolevamer, vancomycin and metronidazole in *Clostridium difficile*-associated diarrhea (CDAD) [abstract K-425a]. In: *Program and abstracts of the 47th Interscience Conference on Antimicrobial Agents and Chemotherapy (Chicago, IL).* Washington, DC: American Society for Microbiology, 2007.

[90] Bouza E, Dryden M, Mohammed R, Peppe J, Chasan-Taber S, Donovan J, Davidson D, Short G. 2008. Results of a phase III trial comparing tolevamer, vancomycin and metronidazole in patients with *Clostridium difficile*-associated diarrhoea, abstr. O464. Programs Abstr. Eighteenth Eur. Congr. Clin. Microbiol. Infect. Dis., Barcelona, Spain. *European Society of Clinical Microbiology and Infectious Diseases*, Basel, Switzerland.

[91] Baines D, Freeman J, Wilcox M. Tolevamer is not efficacious in the neutralization of cytotoxin in a human gut model of *Clostridium difficile* infection. *Antimicrobial agents and chemotherapy* 2009; 53(5): 2202-2204.

[92] Lowy I, Molrine DC, Leav BA, Blair BM, Baxter R, Gerding DN et al. Treatment with monoclonal antibodies against *Clostridium difficile* toxins. *New England Journal of Medicine* 2010; 362(3): 197-205.

[93] Gardiner DF, Rosenberg T, Zaharatos J, Franco D, Ho DD. A DNA vaccine targeting the receptor-binding domain of *Clostridium difficile* toxin A. *Vaccine* 2009; 27(27): 3598-3604.

[94] Sougioultzis S, Kyne L, Drudy D, Keates S, Maroo S, Pothoulakis C, Giannasca PJ, Lee CK, Warny M, Monath TP, Kelly CP. *Clostridium difficile* Toxoid Vaccine in Recurrent *C. difficile*–Associated Diarrhea. *Gastroenterology* 2005; 128: 764-770.

[95] Rubin MS, Bodenstein LE, Kent KC. Severe *Clostridium difficile* colitis. *Dis Colon Rectum.* 1995 Apr;38(4):350-4.

[96] Dallal RM, Harbrecht BG, Boujoukas AJ, Sirio CA, Farkas LM, Lee KK, Simmons RL. Fulminant *Clostridium difficile*: an underappreciated and increasing cause of death and complications. *Ann Surg.* 2002;235(3):363-72.

[97] Impallomeni M, Galletly NP, Wort SJ, Starr JM, Rogers TR. Increased risk of diarrhoea caused by *Clostridium difficile* in elderly patients receiving Cefotaxime. *BMJ* 1995; 311: 1345-6.

[98] Lesna M, Parham DM. Risk of diarrhoea due to *Clostridium difficile* during cefotaxime treatment; Mortality due to *C. difficile* colitis in elderly people has been underestimated. *BMJ* 1996; 312: 778

[99] Mlangeni DA, Harris MD, Franklin L, Hunt P, Karas JA, Enoch DA. Death certificates provide a poor estimate of attributable mortality due to *Clostridium difficile* when compared to a death review panel using defined criteria. *Journal of Hospital Infection* 2011; 77: 370-371.

[100] Karas JA, Enoch DA, Aliyu SH. A review of mortality due to *Clostridium difficile* infection. *Journal of Infection* 2010; 61(1): 1-8.

[101] White HA, Wiselka MJ. Inpatient mortality and death reporting associated with *Clostridium difficile* in a large teaching hospital. *Journal of Hospital Infection* 2011; 77: 369-370.

[102] Oughton MT, Loo VG, Dendukuri N, Fenn S, Libman MD. Hand hygiene with soap and water is superior to alcohol rub and antiseptic wipes for removal of Clostridium difficile. *Infect Control Hosp Epidemiol.* 2009; 30(10: 939-944.

[103] Russell AD. Bacterial resistance to disinfectants: present knowledge and future problems. *J Hosp Infect.* 1999 Dec;43 Suppl:S57-68.

[104] Fawley WN, Underwood S, Freeman J, Baines SD, Saxton K, Stephenson K et al. Efficacy of hospital cleaning agents and germicides against epidemic *Clostridium difficile* strains. *Infection Control and Hospital Epidemiology* 2007; 28(8): 920-925.

[105] Muto CA, Blank MK, Marsh JW, Vergis EM, O'Leary MM, Shutt KA et al. Control of an outbreak with the hypervirulent *Clostridium difficile* BI strain in a University Hospital using a comprehensive "bundle" approach. *Clinical Infectious Diseases* 2007; 45: 1266-1273.

[106] Khan R, Cheesbrough J. Impact of changes in antibiotic policy on *Clostridium difficile*- associated diarrhoea (CDAD) over a five year period in a district general hospital. *Journal of Hospital Infection* 2003; 54: 104-108.

In: Clostridia ISBN: 978-1-62100-761-6
Editors: S. V. Clinton and M. F. Kelly © 2012 Nova Science Publishers, Inc.

Chapter 4

THE LABORATORY DIAGNOSIS OF *CLOSTRIDIUM DIFFICILE* INFECTION

M.R Lekalakala
Department of Medical Microbiology, Faculty of Health Sciences, University of Pretoria and Tshwane Academic Division, National Health Laboratory Service (NHLS), South Africa

ABSTRACT

Clostridium difficile (*C.difficile*) is a gram-positive, spore-forming organism responsible for a spectrum of infections. These infections vary from asymptomatic colonization to severe antibiotic-related diarrhea and life threatening pseudomembranous colitis *C.difficile* forms part of the gastrointestinal tracts flora of most neonates and was believed to be a commensal organism until it was later discovered that *Clostridium difficile* is the primary cause of pseudomembranous colitis [1] and the frequent use of broad-spectrum antibiotics lead to the rapid incidence of *C.difficile* diarrhea [2].

Historically, the cell culture cytotoxicity neutralization assay (CCCN), which detects cytotoxin production in monolayers of cells, such as human diploid fibroblasts, has been the gold standard for *C.difficile* detection in the laboratory but this was labor-intensive hence the evaluation of other methods for the detection of this organism [3]. The laboratory survey which involved forty four laboratories in Ireland revealed that majority of the laboratory used toxin detection by enzyme immunoassay and none of those laboratories used either cell culture or PCR [4].

The aim of this chapter is to review the diagnostic methods available and/or in use to date and to explore the future trends and the role of the laboratory in the diagnosis of *C.difficile* infections.

THE ORGANISM

Clostridium is the genus name of the gram-positive, spore-forming anaerobic bacteria. The genus name comes from the Greek klōstēr (spindle) as the colonies belonging to this resemble spindles used in cloth weaving and long stick with a bulge at the end when viewed under the microscope whereas the species name difficile is a form of the Latin adjective difficilis because, when first identified in 1935 by Hall and O'Toole, the organism was difficult to isolate and grew slowly in pure culture [5, 6, 7, 8]. *C.difficile* is widespread in the soil, water, the intestinal tract of both humans and animals and most hospital surfaces.

Similar to other members of the clostridial family, *C.difficile* is an anaerobic, gram-positive spore-forming rod-shaped bacterium which characteristically forms sub-terminal spores. When cultured on the selective media CCFA (cycloserine-cefoxitin-fructose agar) colonies of *C.difficile* appears flat, yellow, ground-glass with a surrounding yellow halo [8]. The spores are resistant to heat and desiccation and this feature allows long term survival in adverse environments (including hospital environment) and resistance to disinfection, and it contributes to the epidemiology and pathogenesis of the organism [5,7].

PATHOGENICITY

Pathogenic strains of *C.difficile* produce two potent toxins, toxin A, an enterotoxin, and toxin B, a cytotoxin. The genes encoding toxin A and B are encoded in the *C. difficile* pathogenicity locus (*tcdA* and *tcdB*) which also encode two additional regulatory genes (*tcdC* and *tcdD*) [8.9] .The two major toxins, *tcdA* and *tcdB*, have been studied intensively since their initial recognition as major *C. difficile* virulence factors. The results from a collection of studies strongly suggest that detailed analysis of *tcdA* and *tcdB* activity is providing one of the quickest inroads to understanding C. difficile-related disease [8, 9, 10].

Apart from their contribution to disease, *tcdA* and *tcdB* are the primary markers for the diagnosis of C. difficile disease and are detected in the stools of patients. In addition to *tcdA* and *tcdB*, three other open reading frames are located on this pathogenicity locus and are thought to be involved in regulation of toxin production or release [11]. *tcdD* is a proposed positive regulator, *tcdE* is a putative holin protein, and *tcdC* is a proposed negative regulator of toxin gene expression.

Research findings from multiple groups of researchers have revealed a three-domain structure of the large clostridial toxins [11]. The glycosyl-transferase activity is located at the N terminus of the protein, and the C terminus is involved in receptor binding. Located in the middle domain of the protein is a putative transmembrane segment that is thought to be involved in membrane translocation. [11.12]

Even though the primary work on *tcdA* and *tcdB* has been carried out on toxins from the reference strain VPI 10463, it is clear that several genetic variants of these toxins exist in clinical isolates. Some of these toxin variants include hybrids which may have occurred due to genetic exchange between large clostridial toxin-producing clostridia [11]. Some variants are due to subtle sequence variations, deletions, and duplications within the pathogenicity locus account for various toxinotypes of *C. difficile* [11]. Rupnik and colleagues developed a typing system which distinguishes the various toxin types of *C. difficile* [13, 14].Up to 22 toxin types of *C. difficile* containing small sequence variation has been identified [11,15,16].

A newly identified strain of *C. difficile* strains has been characterised, by several typing methods namely: toxinotype III, BI by restriction endonuclease analysis (REA), as North American pulsotype 1(NAP1) by pulsed field gel electrophoresis, and as 027 by PCR. This strain produces 16 times more toxin A and 23 times more toxin B than other strains and also produces a third toxin ,a binary toxin and is also noted to be resistant to the fluoroquinolones: gatifloxacin and moxifloxacin [17, 18, 19]. This strain is commonly referred to as the hypervirulent epidemic *C. difficile* NAP1/B1/027.

The Clinical Presentation of *C.Difficile* Infection(CDI)

C.difficile is the most common cause of infectious diarrhea and antibiotic-associated diarrhea in health care facilities .It may occur as sporadic cases or nosocomial outbreaks [17, 20,21,22,23]. The primary symptoms are watery diarrhea, fever, abdominal pain, and/or tenderness .The infection is usually mild but could in rare cases result in emergency bowel surgery or death [20, 21.24,25] *C.difficile* is becoming both antimicrobial resistant and more virulent, causing greater morbidity [25,26] . Symptoms of CDI are highly variable, ranging from nuisance diarrhea to life-threatening colitis. Typical features include watery diarrhea, with as many as 15–30 bowel movements per day [6, 27]. Most patients complain of abdominal pain or cramps, and they often have lower quadrant tenderness in association with fever and leucocytosis.

Fever, when present, is usually low grade but may involve a temperature as high as 40°C. The peripheral leukocyte count is usually 10,000–20,000 leukocytes/mm3 but may be in the leukemoid range, a feature more commonly recognized now than in the past [28, 29]. Hypoalbuminemia is a common feature because CDI is a protein-losing enteropathy, and some patients may present with anasarca during the late stages of disease [27]. Extra-intestinal symptoms are rare, except for occasional cases of polyarthritis involving large joints [6, 30]. Some patients develop ileus or toxic megacolon; colonic perforation is rare.

The Diagnosis and Laboratory Detection of *C.Difficile* Infection

Unlike most diseases, a simple laboratory assay is not definitive for CDI. The presence of *C. difficile* (either a culture or positive toxin assay) must be confirmed with a clinical presentation of symptoms (usually diarrhea) and other causes of diarrhea must be excluded [27].What complicates the diagnosis of CDI is that some people carry toxin-positive *C. difficile,* yet have no symptoms (asymptomatic carriers) and they may not develop symptoms even after exposure to antibiotics or other typical inciting agents. Conversely, some patients with diarrhea do have negative stool toxin assays.

The cell cytotoxin assay is considered the 'gold standard' for diagnosis but stool culture has the advantage of detecting one-third more cases than cell cytotoxin assay, yet these two methods have lost favor in current practice [28,29]. The elderly, debilitated patients in hospitals and nursing homes are particularly vulnerable. A hypervirulent NAP1, PCR ribotype 027 epidemic strains has been associated with the changing epidemiology and severity of disease [18, 14, 31].

The diagnosis of *C difficile* infection is usually based on the appropriate clinical context, history of recent antibiotic use and diarrhea, along with other symptoms such as fever, abdominal discomfort and/or results from imaging studies in combination with laboratory testing. Although it is critical to consider C difficile infection early on, especially in highly debilitated patients, laboratory testing will be confirmatory.

It is important for the clinicians to recognize the limitations of these diagnostic tests and the need to follow protocols for proper sample selection and handling. When in doubt, the health facility laboratory personnel should be contacted for the latest information and recommendations. There are two categories of laboratory tests for C. difficile: toxin assays (which evaluate for evidence of toxin), and organism detection assays (which evaluate for the presence of the organism. The current cornerstone of laboratory diagnosis is the detection of toxin from faecal samples of which fecal toxin assay using cell cytotoxicity assay (CTA) considered the gold standard [7,8,13,31, 32].

THE LABORATORY DETECTION ASSAYS

1. Cellular Cytotoxicity Assays

The first diagnostic test for *C. difficile* involved neutralization of the cytotoxin by *C. sordellii* or *C.difficile* antitoxin and this study also illustrated that both toxin A and toxin B are cytopathic toxins, but toxin B is approximately 1000-fold more potent in cell lines [33]. The first person with a positive test result was a postoperative patient in California with cephalothin-induced "clindamycin colitis."[7, 21] .This patient had a cytotoxin titer of 10^6, which remains the highest titer measured from the stool of a patient [7, 9].

The initial assay for C. difficile cytotoxicity was done using the cultured mammalian cells [34,35,36] ,and it was through this assay that Gerding et al. [37] showed that 35 (90%) of 39 endoscopy- positive patients were cytotoxin positive. The cellular cytotoxin assay has also been shown to be positive in 97% of all cases of pseudomembranous colitis and in about 27% of patients with antibiotic associated diarrhoea in whom no pseudomembranes were present [38,39,40,41].

Methodology: CTA assay is performed by adding a measured amount of filtered fecal specimen in the presence and absence of specific antiserum to a cultured monolayer of mammalian cells. After 24-48 hours of incubation, the cultured cells are observed for cytopathic effect which occurs as a result of the cell cytoskeleton which appears rounded. Since fecal specimens may contain substances that may induce a nonspecific cytopathic effect, specificity was confirmed with a neutralizing high-titered antiserum [42, 43,44,45,46].

Advantages: The test had the sensitivity and specificity of 98 and 99 respectively when compared with the clinical diagnosis and the laboratory criterion but not the stool culture on the selective media [47, 48]. Its ability to detect other clostridial toxins is also important because no toxin producing strain be missed.

For an optimal bacteriological diagnosis, only liquid stools should be accepted, except in the case of an epidemiological investigation. Due to a rapid loss of cytotoxin activity, only fresh specimens should be processed and stool must be stored at 4 ^{O}C or less, in case tests cannot be performed rapidly. Brazier reported complete inactivation of cytotoxin in about 20% of samples sent through the post in which case they were analysed after few day [49]. On the other hand, cultures of C. difficile remain unaffected by ambient storage due to sporulation.

Repeated samples within7 days of the initial request seem to give little useful information [50] hence repeat testing was not recommended when this method was used for *C.difficile* testing. An advantage of this method was that it is sensitive and specific, and can detect clostridial toxins.

Limitations: The test is relatively slow in comparison to enzyme immunoassays, the nonspecific cytopathic effect (not neutralised by antiserum) which may be observed in about 2% of the cases, rendering any interpretation impossible, the necessity of cell lines maintenance which is usually both time-consuming and expensive, especially if only a small number of specimens are processed.

Of note is that the frequency of cytotoxin detection increases as the severity of symptoms, duration, and pathology increases. The titer tend to increase upon aggravation of symptoms and to decrease with an alleviation of clinical symptoms [52, 53] .

With the advent of new cell culture techniques, the cellular cytotoxicity assay does not require the availability of a CO2 incubator or mammalian cell culture facilities, making the test practical for many smaller hospitals and physicians in rural or remote areas. It employs human foreskin fibroblasts and is performed by inoculating the supplied cell culture with a pre incubated stool filtrate antitoxin mixture [13, 54, 55].

2. Stool Culture for Isolation of C.Difficile

Isolation of *C difficile* organism in stool culture was first proposed by George et.al were they recommended the use of a selective plate containing cycloserine cefoxitin fructose agar (CCFA) [55,56,57,58].This was a major breakthrough in that the same media with or without modification is still available and used by some laboratories for the isolation of *C.difficile* [59,60]. A variety of modifications have been tried to enhance the sensitivity of culture and this included modification of the media itself or pre-treatment of the stool samples [57, 61,62,63]. However this method failed to differentiate toxin-producing from non-toxigenic strains, but permitted molecular typing of the cultured organisms which is essential for monitoring of molecular epidemiology and antibiotic susceptibility [64].

Methodology: Stools are directly inoculated on selective agar plates and incubated in an anaerobic atmosphere for 48 hours. Colonies of *C.difficile* are recognised on culture plates due to their typical morphology (ground glass appearance) and odor (horse manure).A yellow-green fluorescence under ultraviolet illumination is another feature that may be used however confirmation is still essential [64,65]. Gas Liquid Chromatography (GLC) of an agar plug around a suspected colony use to be the best and easiest method for confirming identification because *C. difficile* displays a typical GLC profile [66] with large amounts of butyric and iso-caproic acids.

Due to unavailability of such devices in most laboratories, alternative methods such as individual biochemical tests, anaerobic panels, somatic latex antigen kit and enzyme immunoassays are commonly used. Production of proline-aminopeptidase by a disc test has been proposed as a means of rapid identification of *C.difficile* when used in conjunction with the typical morphology of the colonies [67,68,69].

Advantages: it allows toxin detection, typing and antimicrobial testing of the cultured isolates. It is the only test suitable for epidemiological investigations.

Limitations: The test is labor intensive, long turnaround time (24-48 h) and a low specificity. Due to the existence of toxigenic and non-toxigenic strains, it's apparent that the laboratory needed additional assays for toxin detection on cultures, which implies that culture alone is not sufficient for the diagnosis of *C. difficile* infection.

3. Enzyme Immunoassays (EIA) for C.Difficile Toxin

These are enzymometric immunoassay for the qualitative determination of both *Clostridium difficile* toxins A and B based on polyclonal and monoclonal antibodies against the two toxins. Studies in the early 1980s showed that this assay could detect both toxin A and toxin B but that for toxin B, it was at least 10-fold less sensitive than cell cytotoxicity assays [70,71.72].

EIA are more rapid and easier to perform but are suboptimal if used as stand-alone assays due to the low sensitivity compared to CCNA or toxigenic *C. difficile* culture [1, 8,13,20,24,73,74,75]. Most laboratories rely on *C. difficile* toxin A/B testing alone despite the limits of this approach and have abandoned bacterial culture [74, 76, 77]. Detection of the C. difficile-specific enzyme glutamate dehydrogenase antigen (GDH) may in part replace culture techniques and therefore be an alternative to culture [75,78,79,80].

EIA allows direct detection of Clostridium difficile toxins in stool filtrates [73]. Commercially available assays can detect toxin A alone or both toxins A and B. EIA for both toxins should be used because clinical cases may involve strains with a mutation in toxin A (rendering the EIA for toxin A ineffective) or strains that produce only toxin B [65,82, 83].

The antigen based EIAs (Ag-EIAs) accurately detect an essential and constitutively synthesized enzyme [64, 75], thereby rapidly identifying *C. difficile* while overcoming the low sensitivity of toxin EIAs and suboptimal performance of older, latex-agglutination antigen assays. For well over a decades, first the Premier toxin A enzyme immunoassay (EIA) and then the Premier toxin A and B EIA (A/B EIA; Meridian Bioscience, Inc., Cincinnati, OH) were used to detect this toxin [83,84] with the latter being the most widely used test for the detection of *C. difficile* toxins from stool of patients suspected to have CDI [83].

Some studies [79, 85, 86] have shown that this assay has both high sensitivity of 90% and specificity 95% compared to tissue culture cytotoxicity neutralization assay [1, 51, 74, 78, 85, 87].Although EIAs remain the most frequently used assays among laboratories , enzyme immunoassays (EIAs) for toxin detection lack sensitivity and are considered suboptimal for diagnosing *C. difficile* infection. The two comprehensive manuscripts on EIAs for toxin detection summarized by Carol [88] in her publication, illustrated the extensive work done by both Planche et al. [65] and later by Eastwood and his colleagues [89].

Planche, et al [65] reviewed the literature assessing six toxin A/B tests most commonly used United Kingdom. There were 28 comparisons where a reference method was performed on all tests and those articles were chosen for analysis. In their manuscript, the authors established as acceptability criteria a sensitivity of 90% and a false positive rate of less than or equal to 3% [88]. The assays had to satisfy both criteria to be considered acceptable. Of interest in their findings is that there was significant heterogeneity reported for each individual assay. However when using the diagnostic odds ratio to measure overall accuracy, there was no significant difference in the performance characteristics among the assays evaluated [65, 88]. The authors concluded that toxin EIAs should not be used routinely as a single test for the diagnosis of CDI.

On the other hand Eastwood and colleagues [89] evaluated nine tests namely: six solid phase EIAs and 3 lateral flow assays on the same set of 600 diarrheal stool specimens. The sensitivities ranged from 60% to 81% and the specificities from 91% to 99.4% compared to toxigenic anaerobic culture [65, 89]. The authors assessed the positive predictive values of each test under conditions of low prevalence (2%) in community setting vs. a high prevalence (10%) such as a healthcare setting. The PPV for the low vs. high groups ranged from 32.3% to 68.7%, respectively.

Apart from the variable findings of the evaluation studies done by the two coted authors and others it is clear that most if not all agree that no single EIA is sufficient as a stand-alone test for the diagnosis of C.difficile infections [8,42,65,71,75,88,89]. It is on that basis the most researchers are now looking and the option of either a two-step or three-step algorithms for the diagnosis of *C.difficile* infection.

LATEX AGGLUTINATION TESTS

Knoop et.al [13] in his 1993 review article highlighted some of the controversies of commercial latex agglutination test [13,14].which were reported by Lyerly et al. [90] stating that the latex agglutination test does not detect toxin A but reacts with a latex-reactive protein that contaminates the commercial antigen preparations used to immunize animals for the production of antibody. Several studies have reported on the correlation of the latex agglutination test with cytotoxin testing and clinical disease [13,63,91,92]. However, the published positive and negative predictive values have varied considerably, ranging from 63 to 98%.

Lyerly et al [93] later analysed the same latex agglutination test-reactive protein and the protein was demonstrated to be an enzyme, glutamate dehydrogenase (GDH), which reversibly catalyses the amination of a-ketoglutarate to glutamate and the deamination of glutamate to a-ketoglutarate [94].

Studied over years confirmed that GDH is a common antigen found in toxigenic and non- toxigenic strains of *C.difficile*. It is for this reason that GDH is now used as a marker for the presence of the organism in stool as is therefore recommended as a screening test for *C.difficile* infections.

MOLECULAR ASSAYS

Nucleic acid amplification tests (NAATs) for *C. difficile* were first developed more than 15 years ago, but interest in their clinical use has recently increased as the clinical and epidemiologic importance of CDI has become more widely appreciated [97,98].

NAATs employing both real-time polymerase chain reaction (PCR) and loop mediated isothermal amplification of DNA technologies for the detection of *C. difficile* in stool specimens. The available FDA-approved kits include the GeneOhm (BD), proGastro (Prodesse), GeneXpert (Cepheid), and Illumigene (Meridian) *C. difficile* assays [88].

Most tests are designed to detect the *C. difficile tcdB* gene.

Numerous studies have been published, most in the past several years, detailing the results of controlled evaluations of the performance of NAATs, compared with that of conventional methods for the diagnosis of C. difficile infection [89, 99,100,101]. Despite the variable results from different studies, Overall these assays were found to be generally much more sensitive than are toxin EIAs (90% vs 40%–80%, compared with molecular or culture methods) and demonstrated high specificity [97]. Other studies done by Sharp et.al [101] where they compared the NAAT with GDH and their results were equivalent. On the other hand Weekly et.al [102] showed that NAAT was superior to GDH antigen detection [103]. The variability between studies comparing antigen-based detection to NAATs may be a result of differences in genotype prevalence. Tenover et al [104] have shown that the performance of antigen-based methods fluctuates in a genotype-dependent way, whereas the NAAT included in their study did not exhibit the same variability [103,104]

Advantages: molecular methods provide prompt and accurate diagnosis of *C. difficile* infection when compared to the other non-molecular test. Although the cost of molecular testing is a concern for many laboratories, the rapid detection will enhance early treatment followed by proper infection prevention on control measures thus indirectly reducing the duration of hospitalization and *C. difficile* transmission, along with the need for repeat testing

Limitations: Inhibitory substances in the stool sample may interfere with testing. The test should be done on stool samples of patients who meet the clinical provisional diagnosis of *C.difficile* infection because a PCR positive results in a hospitalized asymptomatic patients may be confusing to the clinician. In very rare cases, the *tcdB* gene target may not be detected by PCR although toxin production is demonstrable by cytotoxin cell assay. Some PCR tests does not differentiate the NAP1 (Ribotype 027) strain from other toxigenic strains of *C. difficile*. Expense inherent to molecular diagnostics may prevent laboratories from utilizing real-time PCR for Clostridium difficile infection. The logistics of molecular diagnostics may also be a limiting factor to clinical microbiology laboratories [105].

USE OF DIAGNOSTIC ALGORITHMS

Testing algorithms are widely used in the laboratory to diagnosis of infectious diseases. Algorithms often involve the use of screening assays which are easily performed, rapid and highly sensitive but sometimes less specific in combination with a highly specific confirmatory. A similar approach have been explore for the diagnosis of *C. difficile*. It is reasonably clear that there is no test that is 100% sensitive and specific for the detection of C. difficile infection. In addition, well studied methods recognized to have high performance characteristics for detection of C. difficile infection are not used routinely in clinical laboratories because of the time, expense and technical expertise required to perform the highly specific and sensitive test.[89, 90]

Ticehurst and colleagues [75] challenged the conventional use of toxin based EIA and ICDs for the diagnosis of *C. difficile* infection. They developed a testing algorithm combining a Glutamate dehydrogenase (GDH) antigen (which is produced by toxigenic and non- toxigenic strains) EIA as a screening test B with a cytotoxin neutralization confirmatory test. Using CTA as a gold standard, they found that toxin A/B EIA had a sensitivity of only 38% far below reported values of 70–90% reported in the literature [85, 93].

Table 1. Assays for the diagnosis of C.difficile cited from Kufelnicka & Kirn [102] (Clinical Infectious Diseases 2011; 52(12):1451–1457)

Assay	Method/ target	Advantages	Disadvantages
Culture	Organism	High sensitivity (often considered to be the Gold standard)	Turn – around time 2-7 days, labor intensive, lacks specificity, does not distinguish between toxigenic and nontoxigenic strain, Isolates must be further tested for the presence of toxin genes (toxigenic cultures).
Cell cytotoxigenic neutralization assay	Functional assay for c. difficile toxin B(TcdB)	Moderate- to- high sensitivity High specificity	48-72 hrs turnaround time , subjective interpretation, requires skilled technicians/technical expertise, labor intensive.
Enzyme immunoassays (EIA) *C.difficile* A (*Tcd*A).	Toxin A detection.	Easy to perform. Rapid turnaround time, inexpensive High specificity.	Low sensitivity Misses TcdA-/TcdB+
EIA TcdB or TcdA/B	Toxin A/B detection	Easy to perform. Rapid turnaround time, inexpensive High specificity	Low specificity and does not distinguish between toxigenic and non-toxigenic strain Positive specimens must be further tested for the presence of toxin(s) or toxin genes.
EIA, glutamate dehydrogenase	Common antigen detection	High sensitivity Good screening test	Low specificity and does not distinguish between toxigenic and nontoxigenic Strains Positive specimens must be further tested for the presence of toxin(s) or toxin genes
Nucleic acid amplification test	Toxin gene(s) detection	High sensitivity and specificity (New gold standard?) Short turn-around time Some easy to perform and minimal hands-on time	Expensive when used to test all samples Detection of asymptomatic colonization a possible concern Some require significant molecular expertise Kufelnicka & Kirn [102] (Clinical Infectious Diseases 2011; 52(12):1451–1457)

In contrast, they found that the GDH antigen test had a sensitivity of 98% and a negative predictive value of 99.7% when cytotoxin neutralization was used as confirmatory test. High sensitivity and negative predictive value are key characteristics of an effective screening test.

In conclusion is clear that accurate diagnosis of *C. difficile* infections is vital for patient management, infection control, and the attainment of reliable surveillance data. The optimal laboratory diagnosis of *C. difficile* infection remains an area of controversy despite the availability of multiple tests.

REFERENCES

[1] Bartlett, J. G. 2006. Narrative review: the new epidemic of *Clostridium difficile*-associated enteric disease. *Annals of Internal. Medicine*. 145:758–764.

[2] Monaghan, T., Boswell T., and Mahida Y.R. 2008. Recent advances in *Clostridium difficile*-associated disease. *Gut* 57:850–860.

[3] Johnson, S., Samore M. H, Shin K. A, B. M., Yoo S.J, and H. J. Oh. 2009. Comparison of two enzyme immunoassays for detection of *Clostridium difficile* toxin A and toxin B.Korean .*Journal of Laboratory Medicine*. 29:122–126>

[4] Fitzpatrick F. , Oza A, Gilleece A ,Byrne A. M, Drudye D .2008. On behalf of the *C. difficile* subcommittee of the Health Protection Surveillance Centre Laboratory diagnosis of *Clostridium difficile* associated disease in the Republic of Ireland. *Journal of Irish microbiology laboratories and Hospital Infection*. 68: 315-321

[5] Emerging Infectious Diseases • www.cdc.gov/eid • Vol. 16, No. 2010

[6] Kelly C P, Pothoulakis C, LaMont J T. 1994. *Clostridium difficile* colitis. *New England Journal of Medicine*.330:257–62

[7] Barlett J.G. 2008. Historical perspectives on the studies of *Clostridium difficile* and *C.difficile* infections. *Clinical Infectious Diseases*. 46:S4–11

[8] Fenner L., Widmer A.F., Goy G., et al. 2008. Rapid and reliable diagnostic algorithm for detection of *Clostridium difficile*. *Journal of Clinical Microbiology*. 46. 328-330.

[9] Bartlett JG, Chang TW, Gurwith M, Gorbach SL, Onderdonk AB. 1978. Antibiotic-associated pseudomembranous colitis due to toxin-producing *Clostridia*. *New England Journal of Medicine*. 298:531–4.

[10] Warny M, Kelly CP. Pathogenicity of Clostridium difficile toxins. In: Hecht G, editor. Microbial Pathogenesis and the *Intestinal Epithelial* Cell. 503 ed. Washington, DC, ASM press, 2003. 502-524 000.

[11] Voth E.D and Ballard. 2005. Clostridium difficile Toxins: Mechanism of Action and Role in Disease. Clinical Microbiology Reviews. 247–263.

[12] Hammond, G. A., and Johnson J.L. 1995. The toxigenic element of Clostridium difficile strain VPI 10463.*Journal of Microbial Pathogenesis*. 19:203–213

[13] Knoop F.C, Owens M, and Crocker C.I.1993. *Clostridium difficile:* Clinical Disease and Diagnosis. *Clinical Microbiology Reviews*. 251-265.

[14] Banno, Y., Kobayashi T,Watanabe T, Ueno K, and Nozawa Y. 1981. Two toxins (D-1 and D-2) of *Clostridium difficile* causing antibiotic-associated colitis: purification an Bartlett JG. Narrative review: the new epidemic of *Clostridium difficile*.

[15] Daniel E. Voth and Jimmy D. Ballard. 2005. *Clostridium difficile* Toxins: Mechanism of Action and Role in Disease. *Clinical microbiology Reviews*. 247–263

[16] Rupnik M., Avesani V., Janc M., Christoph von Eichel-Streiber and Michel Delmée .A Novel Toxinotyping Scheme and Correlation of Toxinotypes with Serogroups of Clostridium difficile Isolates. 1998. *Journal of Clinical Microbiology*. 36: 8, 2240-2247.

[17] Sunenshine R.H., McDonald L.C. 2006. *Clostridium difficile*-associated disease: New challenges from an established pathogen. *Cleveland Clinical Journal of Medicine*. 73. (2): 187-197. Abstract

[18] Warny M., Pepin J., Fang A., et al. 2005. Toxin production by an emerging strain of *Clostridium difficile* associated with outbreaks of severe disease in North American and Europe. *Lancet 366*. 1079-1084. Abstract

[19] McDonald L.C., Killgore G.E., Thompson A., et al. 2005. An epidemic toxin gene variant strain of *Clostridium difficile. New England Journal of Medicine*. 353. 2433-2441. Abstract

[20] Centers for Disease Control and Prevention: *Clostridium difficile infections. http://www.cdc.gov/ncidod/dhqp/id_cdiff_data.html*

[21] Walbrown M.A., Aspinall S.L., Bayliss N.K., et al. 2008. Evaluation *of Clostridium difficile*-associated diarrhea with a drug formulary change in preferred flouroquinolones. *Journal of Manage Care Pharmacy*. 14. (1): 34-40. *Abstract*

[22] Fordtran J.S. 2006. Colitis due to *Clostridium difficile* toxins: Underdiagnosed, highly virulent, and nosocomial. *Proceedings Baylor University Medical Centre.* 19. 3-12.

[23] Halsey J.: Current and future treatment modalities for *Clostridium difficile* associated disease. *American Journal of Health System Pharmacy.* 65. 705-715. *Abstract*

[24] Rupnik M., Kato N., Grabnar M., et al. 2003. New types of toxin A negative, toxin B positive strains among *Clostridium difficile* isolates from Asia. *Journal of Clinical Microbiology.* 41. 1118-1125.2003;

[25] Robin B. McFee, DO, George G.and Abdelsayed, MD. 2009. *Clostridium difficile Disease-a- Month.*55:439-470

[26] Thomas C., Stevenson M., Riley T.2003. Antibiotics and hospital acquired *Clostridium difficile* associated diarrhea: A systemic review. *Journal of Antimicrobial Chemotherapy* 51. (6): 1339-1350.

[27] Barton F J, RW, Alpers DH. Clindamycin-associated colitis. 1974. A prospect study. *Annals of Internal Medicine.* 81:429-33.

[28] Wanahita A, Goldsmith EA, Marino BJ, Musher DM.2003 *Clostridium difficile* infection in patients with unexplained leukocytosis. *America Journal of Medicine.*115:543-6

[29] Reller ME, Lema CA, Perl TM, et al. 2007. Yield of stool culture with isolate toxin testing versus a two-step algorithm including stool toxin testing for detection of toxigenic *Clostridium difficile.* Description of two-step method to detect *C. difficile* using screening step with GDH EIA. *Journal of Clinical Microbiology.* 45:3601-3605.

[30] Putterman C, Rubinow A. 1993. Reactive arthritis associated with *Clostridium difficile* pseudomembranous colitis. *Seminars in rthritis Rheumatism.* 22:420-6

[31] Manabe Y.C, M. Vinetz J.M, Moore R.D, Merz C, Charache P, MD; and. Bartlett J.G, 1995 Clostridium difficile Colitis: An Efficient Clinical Approach to *Ann Intern Med.*;123:835-840

[32] McFarland LV, Clarridge JE, Beneda HW, Raugi GR. 2007. Fluoroquinolone use and risk factors for *Clostridium difficile* disease within a Veterans Administration Healthcare System. *Journal of Clinical Infectious Disease.* 45:1141-1151.

[33] Taylor NS, Thorne GM, Bartlett JG. 1981. Comparison of two toxins produced by *Clostridium difficile. Journal of Infection and Immunity.* 34:1036–43.

[34] Donta, S. T., N. Sullivan, and T. Wilkins. 1982. Differential effects of *Clostridium difficile* toxins on tissue-cultured cells. *Journal of Clinical Microbiology*. 15:1157-1158.

[35] Rihn, B., G. Beck, H. Monteil, F. Lecerf, and R. Girardot.1985. Effect of the cytotoxin of *Clostridium difficile* on cultured hepatoma cells. *Journal of Biology of the Cell*53:23-3

[36] Rothman, S. W., and J. E. Brown. 1981. Inhibition of membrane functions in intact HeLa cells by *Clostridium difficile* cytotoxic culture filtrates. *Current Microbiology*. 6:221- 224.

[37] Gerding, D. N., Olson M. M, Peterson L.R, Teasley D. G,Gebhard R. L, Schwartz M. L, and Lee J. T. 1986. *Clostridium difficile*-associated diarrhea and colitis in adults. *Archives of International Medicine*. 146:95-100

[38] Bartlett, J. G. 1979. Antibiotic-associated pseudomembranous colitis. *Review Infectious Diseases*. 1:530-539.

[39] Bartlett, J. G., T. W. Chang, N. S. Taylor, and A. B. Onderdonk.1979. Colitis induced by *Clostridium difficile*. Review of Infectious Diseases. 1:370-378.

[40] Fekety, R. Prevention and treatment of antibiotic-associated colitis, Washington D C, *American Society for Microbiology*, 1979, p. 276-279.

[41] Sailhamer E.A., Carson K., Chang Y., Zacharias K, Spaniolas K., Tabbara M.,Alam H.B, Moya M.A, Velmahos G, Fulminant. 2009. *Clostridium difficile* Colitis. *Archives of Surgery*. 144: 5.

[42] Viscidi, R. P., B. E. Laughon, M. Hanvanich, J. G. Bartlett, and R. H. Yolken. 1984. Improved enzyme immunoassays for the detection of antigens in fecal specimens. Investigation and correction of interfering factors. *Journal of Immunology*. Methods 67:129-143

[43] Chang, T. W., M. Lauerman, and J. G. Bartlett. 1979. Cytotoxicity assay in antibiotic-associated colitis. *Journal of Infectious Diseases*. 140:765-770.

[44] Bartlett, J. G. 1981. Antimicrobial agents implicated in *Clostridium difficile* toxin-associated diarrhea or colitis. *Journal of Medicine*. 149:6-9.

[45] Rolfe, R. D., and S. M. Finegold. 1979. Purification and characterization of Clostridium difficile toxin. *Infection and Immunity*. 25:191-201.

[46] Florin, I., and M. Thelestam. 1981. Intoxication of cultured human lung fibroblasts with *Clostridium difficile* toxin. *Infection and Immunity*. 33:67-74.

[47] Maniar, A., T. Williams, and G. Hammond. 1987. Detection of *Clostridium difficile* in various tissue culture monolayers. *Journal of Clinical Microbiology*. 25:1999-2000.

[48] Bouzaa E., Peláeza T., Alonsoa R., Catalána P., Muñoza P. and M. Rodríguez M Créixems.2001. 'Second-look' cytotoxicity: an evaluation of culture plus cytotoxin assay of *Clostridium difficile* isolates in the laboratory diagnosis of CDAD *Journal of Hospital Infection* 48: 233-237

[49] Brazier J S. 1993. Role of the laboratory in investigations of *Clostridium difficile* diarrhea. Clinical Infectious Diseases. Suppl.: 4S228–33.

[50] Renshaw AA, Stelling JM, Doolittle MH. 1996. The lack of value of repeated *Clostridium difficile* cytotoxicity assays. *Archives of Pathology and Laboratory Medicine*. 120: 49–52.

[51] Petersen LR, Manson RU, Paule SM, et al. 2007. Detection of toxigenic *Clostridium difficile* in stool samples by real-time polymerase chain reaction for the diagnosis of *C. difficile*-associated diarrhea. *Clinical Infectious Diseases*. 45: 1152-1160

[52] Bartlett, J. G. 1990. *Clostridium difficile*: clinical considerations .*Review Infectious. Diseases*. 12(Suppl.):243-251.

[53] Delmee, M., B. J. Van, A. Simon, M. Janssens, and V. Avesani. 2005. Laboratory diagnosis of Clostridium difficile-associated diarrhoea: a plea for culture. *Journal Medical Microbiology*. 54:187-191

[54] Lyerly, D. M., Krivan H.C, and T. D. Wilkins. 1988. *Clostridium* difficile: its disease and toxins. *Clinical Microbiology*. Rev.1:1-18.

[55] Turgeon D.K, Novicki T.J., Quick J., Carlson L., Miller P., Ulness B., Cent A., Ashley R., Larson a., Coyle m., Limaye A.P., Cookson B.T., Fritsche T.R. 2003 Six Rapid Tests for Direct Detection of Clostridium difficile and Its Toxins in Fecal Samples Compared with the Fibroblast Cytotoxicity Assay . *Journal of clinical microbiology*, p. 667–670

[56] John D. Perry, Kerry Asir, Diane Halimi, Sylvain Orenga, Joanne Dale, Michelle Payne, Ruth Carlton, Jim Evans,and F. Kate. 2010. Gould Evaluation of a Chromogenic Culture Medium for Isolation of *Clostridium difficile* within 24 Hours *Journal of Clinical Microbiology*. 48(11): 3852–3858

[57] Delme´e M. 2001. Laboratory diagnosis of *Clostridium difficile* disease. *Clinical Microbiology Infections*. 7: 411–416

[58] George WL, Sutter VL, Citron D, Finegold SM. 1979. Selective and differential Medium for isolation of *Clostridium difficile*. *Journal of Clinical Microbiology*. 9: 214–9.

[59] Levett PN.1985. Effect of antibiotic concentration in a selective medium on the isolation of *Clostridium difficile* from faecal specimens. *Journal of Clinical Pathology* 38: 233–4.

[60] Peterson L R, Kelly P J, Nordbrock HA. 1996. Role of culture and toxin detection in laboratory testing for diagnosis of *Clostridium difficile* associated diarrhea. European *Journal of Clinical Microbiology and Infectious Diseases*.15: 330–6.

[61] Wilson K H, Kennedy M J, Fekety R F. 1982. Use of sodium taurocholate to enhance spore recovery on a medium selective for *Clostridium difficile*. *Journal of clinical Microbiology*.15: 443–6.

[62] Gerding DN, Brazier JS. 1993 Optimal methods for identifying Clostridium difficile infections . *Clin Infect Dis*. 1993 Jun;16 Suppl 4:S439-42.

[63] Borriello S P, Honour P. 1981. Simplified procedure for the routine isolation of *Clostridium difficile* from faeces. *Journal of clinical pathology*. 34:1124–7.

[64] Wilkins T D, Lyerly D M.2003. *Clostridium difficile* testing: after 20 years, still challenging. *Journal of clinical microbiology*. 41: 531-534

[65] Planche T, Aghaizu A, Holliman R, Riley P, Poloniecki J, Breathnach A, et al. 2008. Diagnosis of *Clostridium difficile* infection by toxin detection kit: A systematic Review. *Lancet infectious diseases*. 8: 777-84.

[66] Brooks, Nunez-Montiel,Basta M T, Basta and Hierholzer J C. 1984. Studies of Stools from Pseudomembranous Colitis, Rotaviral, and Other Diarrheal Syndromes by Frequency-Pulsed Electron Capture Gas-Liquid Chromatography. *Journal of clinical microbiology*. p. 549-560.

[67] Fedorko DP, Williams EC.1997. Use of cycloserine-cefoxitin-fructose agar and L-proline- aminopeptidase (PRO Discs) in the rapid identification of *Clostridium difficile*. *Journal of Clinical Microbiology*. 1997; 35:1258–9.

[68] Rousseau C, Poilane I, Diakite F, Feghoul L, Cruaud P, Collignon A. 2010. Comparison of three *Clostridium difficile* culture media. *Pathologie biologie*. 58 (1): 58-61 abstract

[69] Anne M Anglim and Jens J. Kor. 2006. *C. difficile* - a new epidemic requiring new therapeutics? *Annal internal medicine Intern* published online.

[70] Viscidi R P, Yolken R H, Laughon B E, Bartlett J G.1983. Enzyme immunoassay for detection of antibody to toxins A and B of *Clostridium difficile*. Journal Clinical Microbiology. 18:242–7.

[71] Wren M. W. D, Sivapalan M., Kinson R., and Shetty. P L. 2009. Laboratory diagnosis of *Clostridium difficile* infection.An evaluation of tests for faecal toxin, glutamatedehydrogenase, lactoferrin and toxigenic culturein the diagnostic laboratoryBRITISH. *Journal of Biomedical sciences*. 66 (1)

[72] Goldenberg, S. D., P. R. Cliff, S. Smith, M. Milner, and G. L. French. 2010.Two-step glutamate dehydrogenase antigen real-time polymerase chain reaction assay for detection of toxigenic Clostridium difficile. *J. Hosp. Infect*. 74:48–54.

[73] Gerding D N, Johnson S, Peterson L R, et al. 1985. *Clostridium difficile*-associated diarrhea and colitis. *Journal of Infection Control Hospital and Epidemiology*. 16:459

[74] Connor O, D., Hynes P, Cormican M,Collins E,Corbett-Feeney G, and Cassidy M. 2001. Evaluation of methods for detection of toxins in specimens of feces submitted for diagnosis of *Clostridium difficile*-associated diarrhea. *Journal of Clinical Microbiology*. 39:2846-2849

[75] Ticehurst, J. R., Aird D Z,Dam L M, Borek A P, Hargrove J T, and Carroll K C. 2006. Effective detection of toxigenic *Clostridium difficile* by a two-step algorithm including tests for antigen and cytotoxin. *Journal Clinical. Microbiology*. 44:1145-1149.

[76] Barbut, F., Delmee M, Brazier J S,Petit J C, Poxton I R, Rupnik M, Lalande V, Schneider C, Mastrantonio P,Alonso R, Kuipjer E, and Tvede E. 2003. A European survey of diagnostic methods and testing protocols for *Clostridium difficile*. *Clinical Microbiology Infections*. 9:989-996

[77] Fenner, Andreas F. Widmer, Gisela Goy, Sonja Rudin, and Reno Frei. 2008. Rapid and Reliable Diagnostic Algorithm for Detection of *Journal of Clinical Microbiology*, p. 328-330, Vol. 46, No.

[78] Snell, H., M. Ramos, S. Longo, M. John, and Z. Hussain. 2004. Performance of the Tech Lab *C. DIFF* CHEK-60 enzyme immunoassay (EIA) in combination with the *C. difficile* Tox A/B II EIA kit, the Triage *C. difficile* panel immunoassay, and a cytotoxin assay for diagnosis of *Clostridium difficile*-associated diarrhea. *Journal of Clinical Microbiology*. 42:4863-4865

[79] Van den Berg, R. J., Bruijnesteijn van Coppenraet L S, Gerritsen H-S, Endtz H P, van der Vorm E R, and Kuijper E J. 2005. Prospective multicenter evaluation of a new immunoassay and real-time PCR for rapid diagnosis of *Clostridium difficile*-associated diarrhea in hospitalized patients. *Journal of Clinical Microbiology*. 43:5338-5340

[80] Zheng, L., Keller S F, Lyerly D M, Carman R J, Genheimer C W, Gleaves C A, Kohlhepp S J,Young S,Perez S, and K. Ye. 2004. Multicenter evaluation of a new screening test that detects *Clostridium difficile* in fecal specimens. *Journal Clinical Microbiology*. 42:3837-3840

[81] Knapp C C, Sandin R L, Hall GS, et al. 1993. Comparison of vidas *Clostridium difficile* toxin-A assay and premier *C. difficile* toxin-A assay to cytotoxin-B tissue culture assay for the detection of toxins of *C. difficile*. *Diagnostic Microbiology and Infectious Diseases*. 17:7.

[82] Barbut F, Lalande V, Burghoffer B, et al. 2002. Prevalence and genetic characterization of toxin

[83] Peter H. Gilligan. 2008. Is a Two-Step Glutamate Dehyrogenase Antigen- cytotoxicity Neutralization Assay Algorithm Superior to the Premier Toxin A and B Enzyme Immunoassay for Laboratory Detection of *Clostridium difficile*. *Journal of Clinical Microbiology*. p. 1523–1525

[84] Whittier, S., Shapiro D S, Kelly W F, Walden T P, Wait K J, McMillon L T, and Gilligan P H. 1993. Evaluation of four commercially available enzyme immunoassays for laboratory diagnosis of *Clostridium difficile* associated diseases. *Journal Clinical Microbiology*. 31:2861–2865

[85] Musher, D. M., Manhas A,Jain P, Nuila F,Waqar A,Logan N,Marino B, and Graviss A E. 2007. Detection of *Clostridium difficile* toxin: comparison of enzyme immunoassay results with results obtained by cytotoxicity assay. *Journal of Clinical Microbiology*. 45:2737–2739.

[86] van den Berg, R. J., Vaessen N, Endtz H P, Schulin T, van der Vorm E R ,and Kuijper E J. 2007. Evaluation of real-time PCR and conventional diagnostic methods for the detection of *Clostridium difficile*-associated diarrhoea in a prospective multicentre study. *Journal of Medical Microbiology*. 56:36–42.

[87] Massey, V., Gregson D B, Chagla A H, Storey M, John M A , and Hussain Z. 2003. Clinical usefulness of components of the Triage immunoassay, enzyme immunoassay for toxins A and B, and cytotoxin B tissue culture assay for the diagnosis of *Clostridium difficile* diarrhea. *American Journal of Clinical Pathology*. 119:45–49

[88] Karen C. Carroll. 2011. Tests for the diagnosis of *Clostridium difficile* infection: The next generation,*Anaerobe*. 1 -5

[89] Eastwood K, Else P, Charlett A, Wilcox M. 2009. Comparison of nine commercially available C. *difficile* toxin detection assays, a real-time PCR assay for *C. difficile* tcdB, and a glutamate dehydrogenase

detection assay to cytotoxin testing and cytotoxigenic culture methods. *Journal of Clinical Microbiology*. 47:3211.

[90] Lyerly, D. M., Ball D W, Toth J, and Wilkins T D. 1988. Characterization of cross-reactive proteins detected by Culturette Brand Rapid Latex Test for *Clostridium difficile*. *Journal of Clinical Microbiology*. 26:397-400.

[91] Kelly, M. T., Champagne SG, Shorlock C H, Noble M A, Freeman H J, Smith J A. 1987. Commercial latex agglutination test for detection of *Clostridium difficile*-associated diarrhea. *Journal of Clinical Microbiology*. 25:1244-1247.

[92] Sherman, M. E., DeGirolami P C, Thorne G M, Kimber J, and Eichelberger K. 1988. Evaluation of a latex agglutination test for diagnosis of *Clostridium difficile*-associated colitis. *American Journal of Clinical Pathology*. 89:228-233.

[93] Lyeriy, D. M., Barroso L A, and Wilkins T D. 1991. Identification of the latex test-reactive protein of *Clostridium difficile* as glutamate dehydrogenase. *Journal Clinical Microbiology*. 29:2639-2642.

[94] Isreal Epstein and Nathan Grosswicz. 1975. Purification and Properties of Glutamate Dehydrogenase from a Thermophilic *Bacillus*. *Journal of bacteriology*. 1257-1264

[95] Shahrabadi S, Bryan L E, Gaffney D, Coderre S E,Gordon , Pai C H. 1984. Latex Agglutination Test for Detection of *Clostridium difficile* Toxin in Stool Samples. *Journal of clinical microbiology*. 339-341

[96] Paul D. Stamper, Romina Alcabasa, Deborah Aird, Wisal Babiker, Jennifer Wehrlin,Ijeoma Ikebana, and Karen C. Carroll. 2009.Comparison of a Commercial Real-Time PCR Assay for tcdB Detection to a Cell Culture Cytotoxicity Assay and Toxigenic Culture for Direct Detection of Toxin-Producing *Clostridium difficile* in Clinical Samples. *Journal of Clinical Microbiology*. 373–378

[97] Berth L. Reller , Melvin P. Weinstein, Anna M. Kafelnikov and Thomas J. Kern. 2011. Effective Utilization of Evolving Methods for the Laboratory Diagnosis of *Clostridium difficile* Infection. *Clinical Infectious Disease*. 52 (12): 1451-1457

[98] Kato N, Kato H, et al. 1993. Detection of toxigenic *Clostridium difficile* in stool specimens by the polymerase chain reaction. *Journal Infectious Disease*.167:455-458

[99] Swindells J, Brenwald N, Reading N, Oppenheim B. 2010. Evaluation of diagnostic tests for *Clostridium difficile* infection. *Journal of Clinical Microbiology*. 48:606-608

[100] Larson AM, Fung AM, Fang FC.2010. Evaluation of tcdB real-time PCR in a three-step diagnostic algorithm for detection of toxigenic *Clostridium difficile*. *Journal of Clinical Microbiology.* 48:124-130

[101] Sharp SE, Ruden LO, Pohl JC, Hatcher PA, Jayne LM, Ivie WM.2010. Evaluation of the C.DiffQuikChek Complete Assay, a new glutamate dehydrogenase and A/B toxin combination lateral flow assay for use in rapid, simple diagnosis of *Clostridium difficile* disease. *Journal of Clinical Microbiology*.48:2082-2086.

[102] Anna M. Kufelnicka1 and Thomas J. Kirn ,2011 Effective Utilization of Evolving Methods for theLaboratory Diagnosis of Clostridium difficile Infection. *Clinical Infectious Diseases* ;52(12):1451–1457

[103] Novak-Weekley SM, Marlowe EM, Miller JM, et al. 2010; Clostridium difficile testing in the clinical laboratory by use of multiple testing algorithms. *J Clin Microbiol* ; 48:889–893.

[104] Tenover FC, Novak-Weekley S, Woods CW, et al. 2010 Impact of strain type on detection of toxigenic Clostridium difficile: comparison of molecular diagnostic and enzyme immunoassay approaches. *J Clin Microbiol*; 48:3719–3724

[105] Erik Munson, Dorothy Bilbo, Mary Paul, Maureen Napierala, and Jeanne E. Hryciuk 2011, Modifications of Commercial Toxigenic Clostridium difficile PCR Resulting in Improved Economy and Workflow Efficiency. *JOURNAL OF CLINICAL MICROBIOLOGY*, p. 2279–2282

In: Clostridia
Editors: S. V. Clinton and M. F. Kelly
ISBN: 978-1-62100-761-6
© 2012 Nova Science Publishers, Inc.

Chapter 5

POTENTIAL USE OF *MEGASPHAERA ELSDENII* AS A HYDROGEN PRODUCER

Akihiro Ohnishi*
Department of Fermentation Science,
Faculty of Applied Bio-Science,
Tokyo University of Agriculture,Setagaya-ku,
Tokyo, Japan

ABSTRACT

Over the years, global energy consumption, particularly of fossil fuels, continues to increase. The planet's accessible oil reservoirs have been predicted to be almost depleted within this century. Hydrogen is a promising alternative fuel for sustainable use since the development of alternative energy sources is necessary to establish the long-term sustainability of modern societies. Microbial hydrogen production may reduce the dependence on fossil fuels, reduce carbon dioxide emissions, and unlock unused reserves of bioenergy. Hydrogen fermentation by anaerobic bacteria shows the greatest capacity for microbial hydrogen production. Within the field of hydrogen fermentation research, innovations from a microbiological perspective have been developing continually. In most hydrogen fermentation studies, the genus *Clostridium* has been used as the main hydrogen producer because it appears to produce abundant hydrogen. However, to date, very few industrial systems have progressed beyond the pilot study stage. This

* E-mail address: a1ohnish@nodai.ac.jp

commentary will describe some of the problems associated with the development of industrial-level hydrogen production and will introduce *Megasphaera elsdenii* as one of the microbiological tools currently in development to solve these problems. *M. elsdenii* has some unique characteristics, including its high hydrogen productivity as hydrogen-producing bacteria (HPB) and lactate consumption as lactate-utilizing bacteria (LUB), which will be useful in the design of a practical hydrogen fermentation system.

INTRODUCTION

Hydrogen is a clean and exhaustless energy source; it is also useful as a secondary energy carrier (Winter and Nitsch, 1988). Therefore, techniques for manufacturing hydrogen for practical usage are being studied in various fields. An outline of the main materials and technical theories of hydrogen production are shown in Figure 1 (Ohnishi, in press).

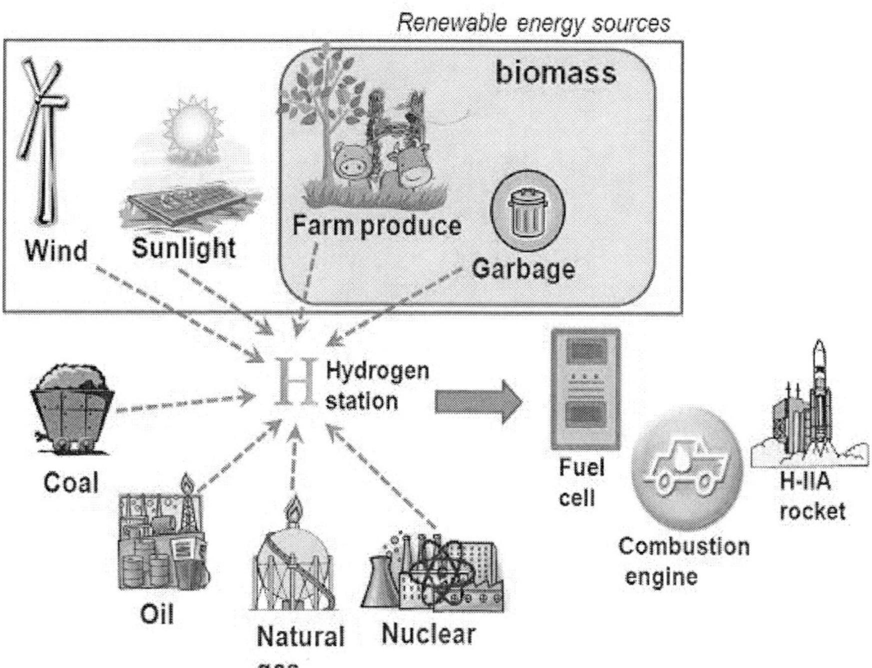

Figure 1. Overview of hydrogen source materials and production processes.

The compounds used as materials for hydrogen production are fossil resources (such as natural gas, oil, and coal), biomass resources (such as organic compounds from animals and plants), and water. The properties are different depending on the compounds, technique, and combinations thereof. Developing the technology to maximize all merits and bypass the deficits is currently underway.

ROLE OF HYDROGEN FERMENTATION

If a fuel such as oil, coal, or natural gas is used as a source material, hydrogen can be extracted effectively by steam reforming and partial oxidation, but the concern of depleting carbon resources would remain. Water is an abundant resource from which hydrogen can be obtained by electrolysis. Nuclear power can be used not only for the electrolysis of water but also in the thermochemical splitting of water in a high-temperature reactor core. Although this offers an advantage for the production of electricity or heat, it also imposes issues of radioactive hazards and accumulation of large quantities of radioactive waste. Pyrolysis gasification and hydrogen fermentation of biomass resources are techniques used to overcome these problems (Claassen et al., 1999; Levin et al., 2004; Kapdan and Kargi, 2006; Ren et al., 2006).

Biomass in various forms exists in abundance in various geographical locations. It is considered the renewable resource that is most carbon neutral and a resource whose collection, storage, and transportation are easy. It is thought that the use of biomass for energy is an important way to contribute to the prevention of global warming, sustainable development formation, strategic industrial creation, rural activation. Biomass can roughly be classified into 2 categories: wet biomass (including garbage, sewage sludge, and livestock excrement) and dry biomass (including paper and wood). Usually pyrolysis gasification is suitable for dry biomass and hydrogen fermentation is suitable for wet biomass. Since hydrogen fermentation can utilize complex waste biomass—comprising ingredients such as food waste as a fermentation substrate—it can maximize from the benefit of low material costs (Li and Fang, 2007). Furthermore, it is feasible for use in small facilities since the hydrogen fermentation process requires temperature conditions of 30–60°C (milder than pyrolysis gasification, which requires the maintenance of thermophilic conditions of more than 600°C). Considering these factors, it may be said that hydrogen fermentation is a technique that can contribute to local energy production for local consumption. Presently, studies in the field of hydrogen fermentation are gaining momentum (Davila-Vazquez et al., 2008; Doi et al., 2009; Ohnishi et al., 2010).

TECHNIQUE OF HYDROGEN FERMENTATION

Various microbes are known to have hydrogen-fermenting abilities(Vos et al., 2009). Despite this, most studies on hydrogen fermentation only use strains from the genus *Clostridium* as the main hydrogen-producing bacteria (HPB), while the study using other microbes is very rare (Iyer et al., 2004). As this reason, it seem only a high hydrogen productivity of the genus *Clostridium* strain at the mesophile region (30-40°C) attracted attention of researchers. However, it is more important that we recognize that they have characteristics suitable for use in the current technique.

In hydrogen fermentation, hydrogen yield is used primarily to evaluate the hydrogen productivity. The hydrogen yield is calculated based on the volume of hydrogen and the substrate that was either added or consumed. This is a good way to compare hydrogen productivity between different HPB. When artificial culture media is use as a substrate, the hydrogen yield is calculated based on the carbon source, such as glucose; it is expressed in mol H_2/mol hexose(Ohnishi et al., 2010). In the case of a complicated substrate, such as waste biomass, it is calculated based on VS volatile solids (VS) or chemical oxygen demand (COD). When an artificial culture media is used as a substrate, yield in the region of 2.4–2.7 mol H_2/mol hexose of hydrogen is obtained by the pure culture of the *Clostridium*. Because in theory up to 4 mol of hydrogen is obtained from 1 mol of glucose in hydrogen fermentation, this genus has around 60% productivity of the theoretical yield(Ren et al., 2006). Since the genus *Enterobacter* (a phylogenetically different HPB) yields a similar amount (2.2–2.3 mol H_2/mol hexose), this is not considered to be a remarkable value. However, the genus *Clostridium* forms "spores," which afford it a greater tolerance for various environmental stresses, including temperature and chemicals(Vos et al., 2009). It is this ability, which in theory makes it very suitable for use in current hydrogen fermentation protocols.

The inhibition of hydrogen production (mainly through substrate competition) and the hydrogen consumption by contaminating saprophytic bacteria decreases hydrogen productivity. Thus, these scenarios must be avoided. In particular, it is known that the presence of lactic acid bacteria (LAB) can cause a decrease in hydrogen productivity through substrate competition and production of lactic acid and antibacterial agents(Noike et al.,

2002). In terms of using waste biomass, there is no technique to prevent these factors, and we must therefore attempt to control it at the system operation level.

In this regard, since spores of the genus *Clostridium* have a high heat tolerance (more than 80°C), heat treatment of the waste biomass (at 90–110°C for around 30 min) and inoculum or thermophilic operation (more than 60°C) of the fermenter using thermophilic strain is very effective as technique (Iyer et al., 2004). Because saprophytic bacteria such as LAB and hydrogen-consuming bacteria generally have a low heat tolerance, heat-treatment and the thermophilic operation of the fermenter sterilizes the saprophytic bacteria and allows the accumulation of *Clostridium* bacteria. Furthermore, because this treatment can also cause a decrease in dissolved oxygen levels, it creates an anaerobic environment suitable for hydrogen fermentation. These techniques are frequently used to stabilize hydrogen yields in current hydrogen fermentation systems, which use the genus *Clostridium* as the main HPB. In past studies, high hydrogen yields were achieved using these treatments alone or in combination. Kim *et al.* (Kim et al., 2008) achieved a high hydrogen yield of up to 80.9 mL H_2/g VS by combining heat treatment (90°C) and base treatment (KOH) of inoculum and using garbage slurry as a substrate. In this manner, the technique improves hydrogen productivity of the hydrogen fermentation and provides stability. However, it is also one of the factors that cannot be accomplished on a scale advantageous to hydrogen fermentation. Because it requires additional facilities, treatment processes, and energy injection, incorporating these improvements might negate the other advantages of the original hydrogen fermentation protocol, such as simplicity, ease, and small scale. For the construction of a hydrogen fermentation system where commercialization is possible, a way around this obstacle in the technical design will be necessary.

MEGASPHAERA ELSDENII AS HPB

I have observed that *Megasphaera elsdenii* demonstrates a high hydrogen fermentation capability when using waste biomass as a substrate. The *M. elsdenii* RG1(Figure 2) strain identified by 16S rDNA analysis (99% similarity) showed a 2.2 ± 0.1 mol H_2/mol hexose hydrogen yield (55% of the theoretical yield) in pure culture(Ohnishi et al., 2010). In addition, it showed high hydrogen productivity when exchanging glucose with lactic acid in the artificial culture media.

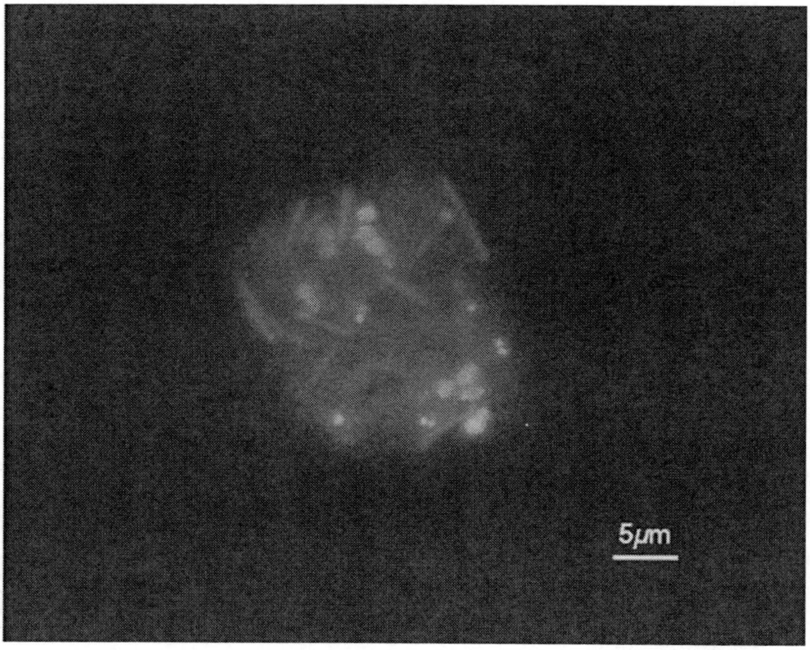

Figure 2. The fermented suspension of two kinds of bacteria after incubation with CFDA in the presence of 0.5mM of EDTA. *M. elsdenii* (cocci) and *Lactobacillus* sp. (rods) cells were visualized by excitation with blue light.

Because the hydrogen yield of *C. butyricum* is approximately 2.4–2.7 mol H_2/mol hexose, the hydrogen productivity of *M. elsdenii* is high enough. However, until now *M. elsdenii* has not been reported as an important HPB. This is perhaps because genus *Megasphaera* seems to comprise non-sporogenic bacteria and that its heat tolerance is low. Even if *M. elsdenii* was included in the inoculum or various biomass types were used as substrates, it would be sterilized along with the other saprophytic bacteria during the heat pretreatment (*Megasphaera* spp. are not spore forming). Furthermore, since the genus *Megasphaera* can utilize lactate, it is possible to omit the LAB sterilization process in the hydrogen fermentation system. The *M. elsdenii* strain RG1 was the main hydrogen producer as part of a microflora mixture, consisting of 8 genera and 13 species (*Atopobium*, *Bifidobacterium*, *Dialister*, *Lactobacillus*, *Megasphaera*, *Prevotella*, *Selenomonas*, and *Veillonella*) and yielded 24.0 ± 1.6 ml H_2/g VS of stable hydrogen in a simple and easy hydrogen fermentation system model using food waste as the substrate without heat pretreatment. The hydrogen yield is lower than previous study; however, the hydrogen fermentation microflora, which included *M. elsdenii* RG1 as the main hydrogen producer, may contribute to the construction of a practical hydrogen fermentation system because it does not require the process of heat pretreatment ahead of hydrogen fermentation.

CONCLUSION

Although *Megasphaera* phylogenetically belongs to Class Clostridia (the gram-positive, low-GC-content group), which includes the genus *Clostridium* frequently used in current hydrogen fermentation protocols, it is a gram-negative genus because of its cell wall composition(Vos et al., 2009). Originally called *Peptostreptococcus elsdenii*, it was reclassified by Rogosa as a new genus, *Megasphaera* (Rogosa, 1971), in the family Veillonellaceae. *M. elsdenii* has been reported to be a normal inhabitant of the gastrointestinal tract in mammals such as humans, cattle, sheep, and pigs (Hino and Kuroda, 1993; Hino et al., 1994; Hashizume et al., 2003). *M. elsdenii* is assumed to be the species primarily responsible for lactate utilization in the rumen and in the large intestine of pigs and rodents (Tsukahara et al., 2002). In humans, some cell culture studies have reported *M. elsdenii* to be present in feces (Hayashi et al., 2002). The possibility of using such an HPB to produce bioenergy casts a new spotlight over the relationship between humans and *M. elsdenii.*

REFERENCES

Claassen, P., Van Lier, J., Lopez Contreras, A., Van Niel, E., Sijtsma, L., Stams, A. et al. (1999) Utilisation of biomass for the supply of energy carriers. *Applied Microbiology and Biotechnology* 52: 741-755.

Davila-Vazquez, G., Arriaga, S., Alatriste-Mondragon, F., de Le n-Rodr guez, A., Rosales-Colunga, L., and Razo-Flores, E. (2008) Fermentative biohydrogen production: trends and perspectives. *Reviews in Environmental Science and Biotechnology* 7: 27-45.

Doi, T., Matsumoto, H., Abe, J., and Morita, S. (2009) Feasibility study on the application of rhizosphere microflora of rice for the biohydrogen production from wasted bread. *Int. J. Hydrogen Energy* 34: 1735-1743.

Hashizume, K., Tsukahara, T., Yamada, K., Koyama, H., and Ushida, K. (2003) *Megasphaera elsdenii* JCM1772T normalizes hyperlactate production in the large intestine of fructooligosaccharide-fed rats by stimulating butyrate production. *Journal of Nutrition* 133: 3187.

Hayashi, H., Sakamoto, M., and Benno, Y. (2002) Phylogenetic analysis of the human gut microbiota using 16S rDNA clone libraries and strictly anaerobic culture-based methods. *Microbiology and immunology* 46: 535-548.

Hino, T., and Kuroda, S. (1993) Presence of lactate dehydrogenase and lactate racemase in *Megasphaera elsdenii* grown on glucose or lactate. *Applied and Environmental Microbiology* 59: 255.

Hino, T., Shimada, K., and Maruyama, T. (1994) Substrate preference in a strain of *Megasphaera elsdenii*, a ruminal bacterium, and its implications in propionate production and growth competition. *Applied and Environmental Microbiology* 60: 1827.

Iyer, P., Bruns, M., Zhang, H., Van Ginkel, S., and Logan, B. (2004) H_2-producing bacterial communities from a heat-treated soil inoculum. *Applied Microbiology and Biotechnology* 66: 166-173.

Kapdan, I., and Kargi, F. (2006) Bio-hydrogen production from waste materials. *Enzyme and Microbial Technology* 38: 569-582.

Kim, S.H., Han, S.K., and Shin, H.S. (2008) Optimization of continuous hydrogen fermentation of food waste as a function of solids retention time independent of hydraulic retention time. *Process Biochemistry* 43: 213-218.

Levin, D., Pitt, L., and Love, M. (2004) Biohydrogen production: prospects and limitations to practical application. *International Journal of Hydrogen Energy* 29: 173-185.

Li, C., and Fang, H. (2007) Fermentative hydrogen production from wastewater and solid wastes by mixed cultures. *Critical Reviews in Environmental Science and Technology* 37: 1-39.

Noike, T., Takabatake, H., Mizuno, O., and Ohba, M. (2002) Inhibition of hydrogen fermentation of organic wastes by lactic acid bacteria. *International Journal of Hydrogen Energy* 27: 1367-1371.

Ohnishi, A. (in press) Search and evaluation method of hydrogen fermentation microflora. Hydrogen production: Prospects and Processes. *Nova Science*.

Ohnishi, A., Bando, Y., Fujimoto, N., and Suzuki, M. (2010) Development of a simple bio-hydrogen production system through dark fermentation by using unique microflora. *International Journal of Hydrogen Energy* 35: 8544-8553.

Ren, N., Li, Y., Wang, A., Li, J., Ding, J., and Zadsar, M. (2006) Hydrogen production by fermentation: review of a new approach to environmentally safe energy production. *Aquatic Ecosystem Health and Management* 9: 39-42.

Rogosa, M. (1971) Transfer of *Peptostreptococcus elsdenii* Gutierrez et al. to a new genus, *Megasphaera* [*M. elsdenii* (Gutierrez et al.) comb. nov.]. *International Journal of Systematic and Evolutionary Microbiology* 21: 187.

Tsukahara, T., Koyama, H., Okada, M., and Ushida, K. (2002) Stimulation of butyrate production by gluconic acid in batch culture of pig cecal digesta and identification of butyrate-producing bacteria. *Journal of Nutrition* 132: 2229.

Vos, P., Garrity, G., Jones, D., Krieg, N., Ludwig, W., Rainey, F. et al. (2009) *Bergey's Manual of Systematic Bacteriology* 2nd edn, vol. 3: The Firmicutes.

Winter, C., and Nitsch, J. (1988) *Hydrogen as an energy carrier.*

INDEX

B

C

D

F

G

N

O

P

Q

R

S

T